CURRENT INTERRUPTION TRANSIENTS CALCULATION

CURRENT INTERRUPTION TRANSIENTS CALCULATION

David F. Peelo
Consultant, former Specialist Engineer at BC Hydro, Vancouver, Canada

WILEY

This edition first published 2014
© 2014 John Wiley and Sons Ltd.

Registered office
John Wiley & Sons Ltd, The Atrium, Southern Gate, Chichester, West Sussex, PO19 8SQ, United Kingdom

For details of our global editorial offices, for customer services and for information about how to apply for permission to reuse the copyright material in this book please see our website at www.wiley.com.

The right of the author to be identified as the author of this work has been asserted in accordance with the Copyright, Designs and Patents Act 1988.

All rights reserved. No part of this publication may be reproduced, stored in a retrieval system, or transmitted, in any form or by any means, electronic, mechanical, photocopying, recording or otherwise, except as permitted by the UK Copyright, Designs and Patents Act 1988, without the prior permission of the publisher.

Wiley also publishes its books in a variety of electronic formats. Some content that appears in print may not be available in electronic books.

Designations used by companies to distinguish their products are often claimed as trademarks. All brand names and product names used in this book are trade names, service marks, trademarks or registered trademarks of their respective owners. The publisher is not associated with any product or vendor mentioned in this book.

Limit of Liability/Disclaimer of Warranty: While the publisher and author have used their best efforts in preparing this book, they make no representations or warranties with respect to the accuracy or completeness of the contents of this book and specifically disclaim any implied warranties of merchantability or fitness for a particular purpose. It is sold on the understanding that the publisher is not engaged in rendering professional services and neither the publisher nor the author shall be liable for damages arising herefrom. If professional advice or other expert assistance is required, the services of a competent professional should be sought.

Library of Congress Cataloging-in-Publication Data

Peelo, David F.
  Current interruption transients calculation / by David F. Peelo.
      1 online resource.
  Includes bibliographical references and index.
  Description based on print version record and CIP data provided by
publisher; resource not viewed.
      ISBN 978-1-118-70719-7 (Adobe PDF) – ISBN 978-1-118-70721-0 (ePub) – ISBN
978-1-118-60047-4 (cloth)   1. Transients (Electricity)–Mathematical models.
  I. Title.
     TK3226
     621.319'21–dc23
                                                                2013038584

A catalogue record for this book is available from the British Library.

ISBN: 978-1-118-60047-4

Set in 10/12 pt TimesLTStd-Roman by Thomson Digital, Noida, India
Printed and bound in Singapore by Markono Print Media Pte Ltd

1   2014

Contents

Preface   ix

1 Introduction   **1**
  1.1 Background   1
  1.2 Short-Circuit Rating Basis for High-Voltage Circuit Breakers   2
  1.3 Current Interruption Terminology   4
  Bibliography   7

2 RLC Circuits   **8**
  2.1 General   8
  2.2 Series RLC Circuit with Step Voltage Injection   8
  2.3 Source-Free Series RLC Circuit with Precharged Capacitor   15
  2.4 Source-Free Parallel RLC Circuit with Precharged Capacitor   18
  2.5 Parallel RLC Circuit with Ramp Current Injection   21
  2.6 Alternative Equations   27
  2.7 Travelling Wave Basics   28
  2.8 Summary   32
  Bibliography   32

3 Pole Factor Calculation   **33**
  3.1 General   33
  3.2 Pole Factors: Effectively Earthed Systems   43
  3.3 Pole Factors: Non-Effectively Earthed Systems   51
  3.4 Alternative Pole Factor Calculation Method   56
  3.5 Three-Phase Test Circuit Arrangement   57
  3.6 Summary   59
  Bibliography   60

4 Terminal Faults   **61**
  4.1 General Considerations   61
  4.2 Standard TRV Derivation   63
  4.3 Effect of Added Capacitance   72
  4.4 Effect of Added Resistance   83

4.5	Effect of Added Inductance	84
4.6	Out-of-Phase Switching	90
4.7	Asymmetrical Currents	91
4.8	Double Earth Faults	99
4.9	Summary	102
	Bibliography	103

5 Short-Line Faults — 104

5.1	General	104
5.2	Line-Side Voltage Calculation	104
5.3	Effect of Added Capacitance	115
5.4	Discussion	118
	Bibliography	119

6 Inductive Load Switching — 120

6.1	General	120
6.2	General Shunt Reactor Switching Case	123
6.3	Shunt Reactors with Isolated Neutrals	130
6.4	Shunt Reactors with Neutral Reactor Earthed Neutrals	135
6.5	Shunt Reactors with Earthed Neutrals	136
6.6	Re-Ignitions	137
6.7	Unloaded Transformer Switching	139
6.8	Discussion	139
6.9	Summary	139
	Bibliography	140

7 Capacitive Load Switching — 141

7.1	General	141
7.2	Shunt Capacitor Banks	141
	7.2.1 Energization	141
	7.2.2 De-Energization	152
	7.2.3 Outrush	159
7.3	Transmission Lines	160
7.4	Cables	163
7.5	Summary	165
	Bibliography	166

8 Circuit Breaker Type Testing — 167

8.1	Introduction	167
8.2	Circuit Breaker Interrupting Time	167
8.3	Inherent Transient Recovery Voltages	173
8.4	Inductive Load Switching	174
8.5	Capacitive Current Switching	175
	Bibliography	175

Appendix A: Differential Equations	**177**
Bibliography	186
Appendix B: Principle of Duality	**187**
Appendix C: Useful Formulae	**190**
Appendix D: Euler's Formula	**193**
Bibliography	196
Appendix E: Asymmetrical Current-Calculating Areas Under Curves	**197**
Appendix F: Shunt Reactor Switching: First-Pole-to-Clear Circuit Representation	**200**
Appendix G: Special Case: Interrupting Small Capacitive Currents	**207**
Bibliography	210
Appendix H: Evolution of Transient Recovery Voltages	**211**
H.1 Introduction	211
H.2 TRVs: Terminal Faults	212
H.3 Terminal Fault TRV Standardization	218
H.4 Short-Line Fault	220
H.5 Inductive and Capacitive Load Current Switching	221
H.6 Terminal Fault TRV Calculation	221
H.6.1 Pole Factor Calculation	221
H.6.2 Transient Calculation	225
Bibliography	226
Index	**231**

Preface

After a fortunate and rewarding career that started at ASEA in Ludvika, Sweden, and was followed by 28 years at BC Hydro in Vancouver, Canada, I took early retirement in May 2001. Not long afterward, I was asked by the Association of Professional Engineers and Geoscientists of British Columbia (B.C.) if I would be interested in presenting continuing professional development courses on circuit breaker application, and this started a second career in teaching.

The first course was 4 hours long and eventually grew into much more detailed courses, some up to 5 days' duration. Experience with the courses showed that the part that generated the most questions from participants related to all types of current interruption transients, and I started to consider developing a course on transients alone. At about the same time, the engineering manager at one of my consulting clients lamented the fact that engineers today, particularly the younger generation of engineers, are much too dependent on software and have lost sight of theory and practical reality. He asked if a course could be developed to provide a fundamental understanding of transients and enable estimations using only a hand calculator and a spreadsheet program.

The approach taken (after a number of false starts) was to draw the circuit diagrams for all possible making, breaking, reignition and restriking cases. Comparison showed that practically all cases are covered by four basic circuits (Tables 2.1 and 2.2). Some exceptions, of course, occur but are variations on a common theme. Three of the circuits involve second-order linear homogeneous differential equations that, instead of individually resorting to Laplace transformation-based solutions, have a common solution of the form

$$y = A\,e^{r_1 x} + B\,e^{r_2 x},$$

where the roots r_1 and r_2 are derived from the circuit RLC components and the constants A and B from the initial or boundary conditions. The equation in turn has three possible variations: the roots are real, equal or complex corresponding to overdamping, critical damping and underdamping, respectively. After being derived, the three equations enable a generic approach to RLC oscillatory circuit calculations (Table A.1).

The fourth case involves a second-order non-homogeneous differential equation that is more difficult to solve than the homogeneous case. However, mathematicians have long resolved the difficulty by providing lookup tables, basically making a guess at the solution, and then using the method of undetermined coefficients to solve the equation (Appendix A).

At this stage, we now have three equations for each of the four circuits incorporating the r_1 and r_2 roots. The next step is to apply the boundary conditions, and the equations for current or voltage in real time are derived. The final step is to convert the equations to a generic format by expressing the circuit damping and time in relative terms, that is, damping relative to critical damping and time relative to the period of the frequency of the transient oscillation (Tables 2.1 and 2.2). General curves can then be drawn and are easily convertible to current or voltage in real time for any switching case.

For multiphase faults, sequential interruption of the fault current in the individual circuit breaker poles leads to AC recovery voltages higher than rated voltage. The AC recovery voltages are related to rated voltage—actually prefault voltage at the point of the fault—by pole factors calculated using the method of symmetrical components. A number of approaches are considered, including a generic approach to first-pole-to-clear pole factor calculation.

As readers will learn, there is a certain symmetry to current interruption transients. For any switching event, taking first the status before the switching operation and then the status after the operation, the transient is the transition from "before" to "after." On this basis, all transients have a starting point, an aiming point or axis of oscillation and a maximum point. Take, for example, the transient recovery voltage (TRV) for a terminal fault on an effectively earthed system: the starting point is zero, the axis of oscillation is around the AC recovery voltage and the maximum value is dependent on the damping and nature of the involved circuit. Understanding this overall concept enables a graphical approach to transient calculation in many cases (see Figures 6.7 and 6.8).

This is not a book about circuit breaker application, and readers are referred in this regard to the Bibliography in Chapter 1. Also, it is not a book about how to use Excel for equation-based calculations; guidance is readily available in instruction manuals and online. Using the generic approach to transient calculation is well suited to Excel because generic time is always in radians, a prerequisite for plotting sinusoidal and hyperbolic functions. A note of caution with respect to plotting in Excel is that, in contrast to software that permits the plotting of functions, Excel plots points. This means, for example, if no point is calculated at a maximum value, then the maximum value will not appear in the plot. A further note is in combining plots with different frequencies, such as the case of adding series reactors, all plots have to be referred to common real-time coordinates before attempting addition or subtraction.

The book is intended to be inclusive. The switching cases are covered in detail in the main text, and supporting calculations and information can be found in Appendices A–G. The evolution of TRVs and their understanding is interesting and is reviewed in Appendix H. The first circuit breakers became commercially available around 1910, and technical papers started to appear within a few years in AIEE publications. The notion of a TRV was first recognized in 1927 by J.D. Hilliard of GE, who used the descriptive term "voltage kick" for the concept. The first standards for fault current TRVs were developed in the 1950s and evolved further into the standards of today.

I would not have been able to write this book without the support that made my career possible. I am grateful to BC Hydro for supporting my participation in learned societies, principally CIGRE and the IEEE, and in the development of circuit breaker standards in IEC; to my colleagues past and present at BC Hydro and in the IEEE Switchgear Committee, CIGRE Study Committee A3 and IEC Technical Committee 17A; to those who have attended the course and asked the great questions that contributed to the book content; and, most of all, to

Sandra Giasson for her patient and diligent word processing of the text through several drafts to the final version.

Writing this book took 10 months, but really it has been 30 years in the making. Now it's done, and I hope that you will find it to be useful and of value.

David F. Peelo
Vancouver, B.C., Canada

1

Introduction

1.1 Background

The intent of this textbook is to explain the origin and nature of the transients associated with fault and inductive and capacitive load current interruption. The transients in general have a power frequency and an oscillatory component. The oscillatory components have an RLC circuit basis with such a degree of commonality between the above current interruption cases that a generic calculation approach is possible. The power frequency component is either a balanced or momentarily unbalanced quantity and in some cases is the axis of oscillation for the oscillatory component. In overview, the following transients will be analyzed and the resulting equations applied to real current interruption cases:

- *Fault current interruption*: The transient of interest is the transient recovery voltage (TRV) that appears across the circuit breaker after current interruption. For terminal faults, that is, a fault at the circuit breaker, the power frequency component is dependent on the system earthing and the type of fault. The oscillatory component can be either overdamped or underdamped with travelling waves contributing to the former oscillation. The TRV may be on one side of the circuit breaker only, for example, a three-phase-to-earth fault on an effectively earthed system, or on both sides of the circuit breaker as for the out-of-phase switching and short-line fault cases.
- *Inductive current interruption*: The transients for consideration in this case are the TRV, which is the difference between the source power frequency and the load circuit oscillation, as well as the transients due to re-ignitions. The load circuit and re-ignition transient oscillations are underdamped.
- *Capacitive current interruption*: The transients in this case are related to both current and voltage. The transient currents to be considered are those due to inrush on switching in a single shunt capacitor bank or in back-to-back switching and outrush current when a bank discharges into a fault. At current interruption, the TRV is the difference between the source power frequency voltage and the trapped DC voltage on the capacitive load at current interruption. The voltage transient of issue is that due to re-striking.

The structure of the textbook is the following:

- Chapter 1: The short-circuit rating basis for high-voltage circuit breakers is described with reference to the IEC circuit breaker standard IEC 62271-100 followed by a review of current interruption terminology.
- Chapter 2: Oscillatory RLC circuits are treated using a generic solution approach without any recourse to the traditional Laplace transform method. An examination of all the various circuits involved in current interruption, re-ignitions or re-striking and making showed that by treating four basic circuit configurations, almost all switching cases can be covered. Some exceptions, of course, occur but, as readers will appreciate later, these are actually variations on a common theme. A basic knowledge of travelling waves is required later in the text, and an overview of basic considerations is included in this chapter.
- Chapter 3: Symmetrical component theory is applied to calculate the unbalanced power frequency voltage values, expressed as per-unit pole factors, that occur during fault and inductive and capacitive load current interruption.
- Chapter 4: The basis for the TRVs for terminal faults, that is, faults located at the terminals of the circuit breaker, is derived. This basis is then applied to the test duties required by IEC 62271-100 and further to show the effects of added capacitance, opening resistors and series reactors. The special cases of out-of-phase switching and double earth faults are then treated. This is followed by the derivation of asymmetrical current requirements and the relationship to time constants and so-called X/R values.
- Chapter 5: The short-line fault is a special case with the circuit breaker being stressed by the difference between TRVs on the source and line sides. The derivation of the line-side transient, which is not oscillatory in the usual sense but rather is travelling wave based, is described and related to standard requirements.
- Chapter 6: Inductive load current switching includes the switching out of unloaded transformers and shunt reactors. The former switching case is not onerous for circuit breakers but the same cannot be said for the latter case. The multiple variations of shunt reactor switching configurations are treated using the generic approach.
- Chapter 7: Capacitive load current switching involves both the switching in and switching out of shunt capacitor banks and unloaded cables and lines. The derivation of inrush and outrush currents, TRVs and re-striking events are treated in detail.
- Chapter 8: Circuit breaker type testing requirements for fault current interruption and load current switching are reviewed.

No chapter is stand-alone as such, and readers should note that Chapters 2 and 3 provide the basic theory for the calculations of Chapters 4–7.

Supporting calculations and information relating to the main text are provided in Appendices A–G. Finally, a brief history of how the understanding and appreciation of TRVs evolved and became standards is provided in Appendix H.

1.2 Short-Circuit Rating Basis for High-Voltage Circuit Breakers

High-voltage circuit breakers are rated on the basis of clearing three-phase faults. The most onerous case with respect to TRVs is for the first-pole-to-clear (FPTC or fptc). This is because after the first circuit breaker pole clears, the system becomes unbalanced, causing the AC recovery voltage across the pole to exceed its normal phase-to-earth value. Two cases can be distinguished based on the earthing of the power system:

Case 1: Power system effectively earthed.
  An effectively earthed power system is one in which the ratio of the zero-sequence reactance to the positive-sequence reactance is positive and equal to 3 or less (neutrals solidly or low impedance earthed). Circuit breakers applied on such systems are rated on the basis of clearing a three-phase-to-earth fault. After the most onerous first pole clearing, this leaves a double-phase-to-earth fault and, in turn, after second pole clearing, leaves a single-phase-to-earth fault to be cleared by the third pole. This sequence is shown in Figure 3.6.

Case 2: Power system non-effectively earthed.
  A non-effectively earthed power system is not defined by sequence reactances but rather as one where the neutral is isolated, high impedance or resonant earthed. Circuit breakers applied on such systems are rated on the basis of clearing a three-phase unearthed fault. First pole clearing leaves a phase-to-phase fault to be cleared simultaneously by the second and third poles in series. Before the second and third pole clearing, the fault-side neutral will shift by 0.5 pu and the AC recovery voltage for the first-pole-to-clear is 1.5 pu. This sequence is shown in Figure 3.7.

The standard TRV requirements for a 245 kV circuit breaker on an effectively earthed system and a 72.5 kV circuit breaker on a non-effectively earthed system are given in Tables 1.1 and 1.2, respectively.

Without going into detail at this point, the TRVs are based on the two components briefly discussed earlier: a power frequency component given by the first-pole-to-clear factor k_{pp} (Chapter 3) and an oscillatory component, which may actually be aperiodic, given by the amplitude factor k_{af} (Chapter 2).

The short-line fault and out-of-phase switching requirements are also shown in Tables 1.1 and 1.2.

In general, circuit breakers are designed to withstand voltage, carry load current and clear faults. However, circuit breakers are also required to interrupt load currents. Load currents at or around unity power factor present no difficulty, but at zero power factor leading or lagging,

Table 1.1 Standard transient recovery voltage values for 245 kV rated circuit breaker on an effectively earthed system.

Rated voltage, U_r (kV)	Test duty	First-pole-to-clear factor, k_{pp} (pu)	Amplitude factor, k_{af} (pu)	First reference voltage, u_1 (kV)	Time, t_1 (μs)	TRV peak value, u_c (kV)	Time, t_2 (μs)	Time delay, t_d (μs)	Voltage, u' (kV)	Time, t' (μs)	RRRV, u_1/t_1 (kV/μs)
245	Terminal fault	1.3	1.40	195	98	364	392	2	98	51	2
	Short-line fault	1	1.40	150	75	280	300	2	75	40	2
	Out-of-phase	2	1.25	300	196	500	392–784	2–20	150	117	1.54

RRRV: rate of rise of recovery voltage.

Table 1.2 Standard transient recovery voltage values for 72.5 kV rated circuit breaker on a non-effectively earthed system.

Rated voltage, U_r (kV)	Type of test	First-pole-to-clear factor, k_{pp} (pu)	Amplitude factor, k_{af} (pu)	TRV peak value, u_c (kV)	Time, t_3 (µs)	Time delay, t_d (µs)	Voltage, u' (kV)	Time, t' (µs)	RRRV, u_c/t_3 (kV/µs)
72.5	Terminal fault	1.5	1.54	137	93	5	45.6	36	1.47
	Short-line fault	1	1.54	91.2	93	5	30.4	36	0.98
	Out-of-phase	2.5	1.25	185	186	28	61.7	90	0.99

current interruption is an onerous duty. No rated interrupting current values are stated in the standards because in practice they are application dependent. Preferred capacitive current switching ratings are stated in the expectation that type testing to these values will cover a majority of actual applications.

1.3 Current Interruption Terminology

Current interruption terminology can be understood by considering an actual event. Figures 1.1–1.3 show the trace of a close–open (CO) three-phase unearthed fault current test on a vacuum circuit breaker. Taking each figure in turn, the terminology is as follows:

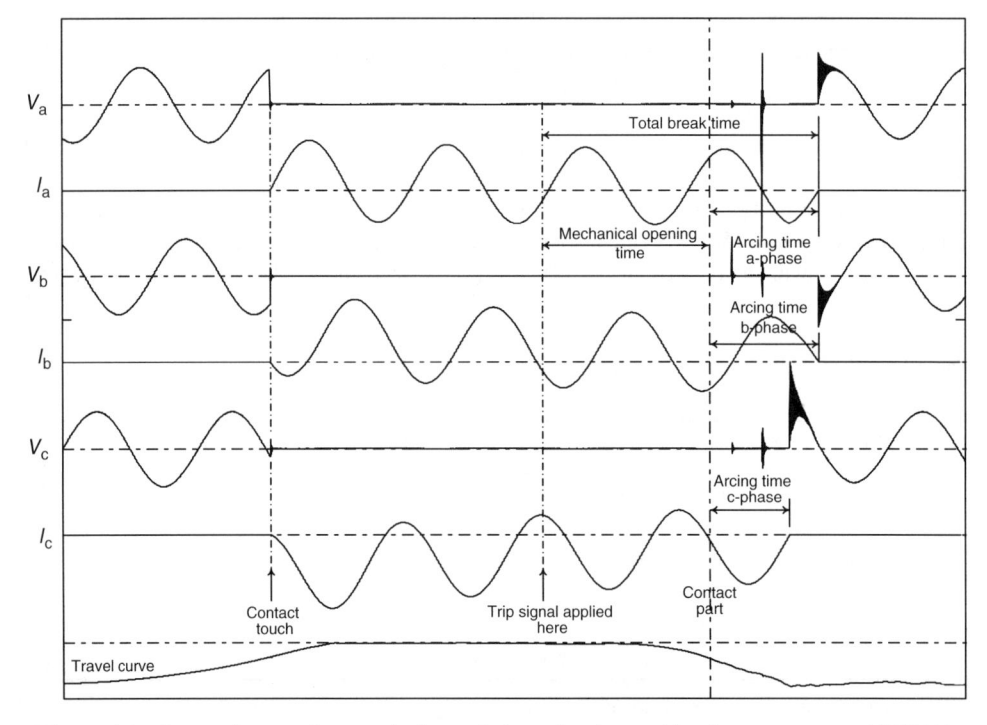

Figure 1.1 Current interruption terminology: timing-related quantities (trace courtesy of KEMA).

Introduction 5

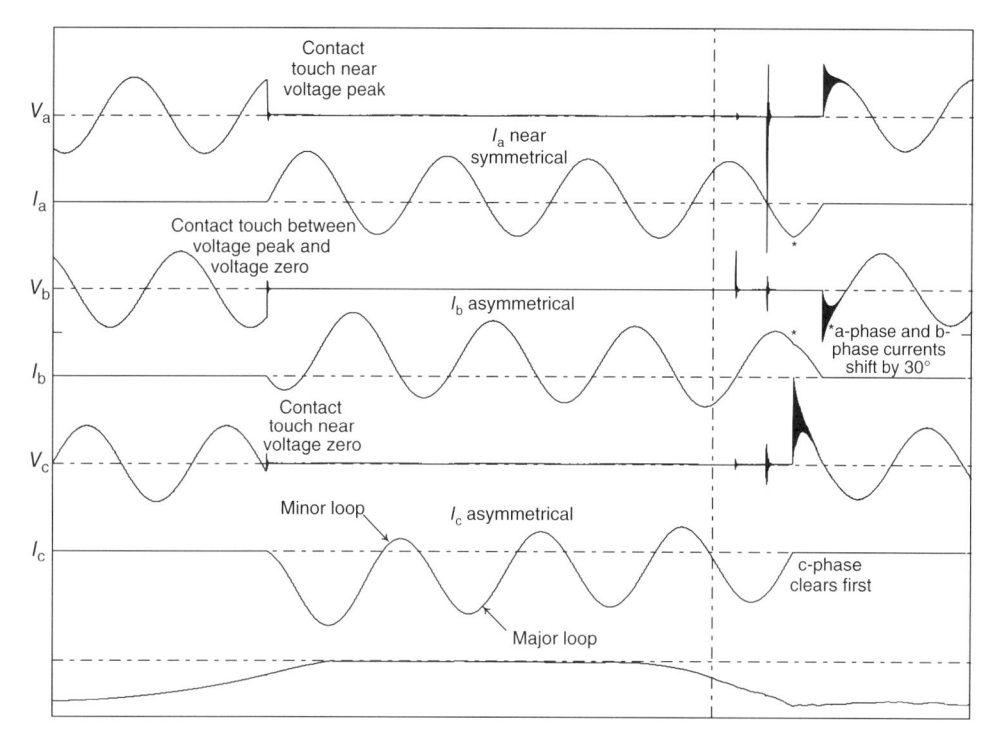

Figure 1.2 Current interruption terminology: current-related quantities (trace courtesy of KEMA).

Figure 1.1 (timing-related quantities):

- The circuit breaker is initially open, and a close signal is applied to the close coil to initiate closing.
- After a short electrical delay time, the moving contact starts in motion (travel curve at the bottom of the trace) and makes contact with the circuit breaker fixed contact. This instant is referred to as contact touch or contact make. In practice, actual electrical making of the circuit may precede mechanical contact because of a prestrike between the contacts. The time between application of the close signal and contact touch is the mechanical closing time of the circuit breaker.
- The circuit breaker is now closed and carrying fault current. A trip signal is applied to the trip coil initiating opening, also referred to as tripping, of the circuit breaker. After a short electrical time delay, the moving contact is set in motion and mechanical separation of the fixed and moving contacts occurs. This instant is referred to as contact part, contact parting or contact separation. The time between application of the trip signal and contact part is the mechanical opening time.
- An arc is drawn between the contacts, and current interruption attempts are made as the zero crossings occur, first on b-phase, then on a-phase and successfully on c-phase. c-phase is thus the first-pole-to-clear with an arcing time—time between contact part and current interruption—of about one half-cycle. The interrupting time, also referred to as the break time, on c-phase is the mechanical opening time plus the arcing time.

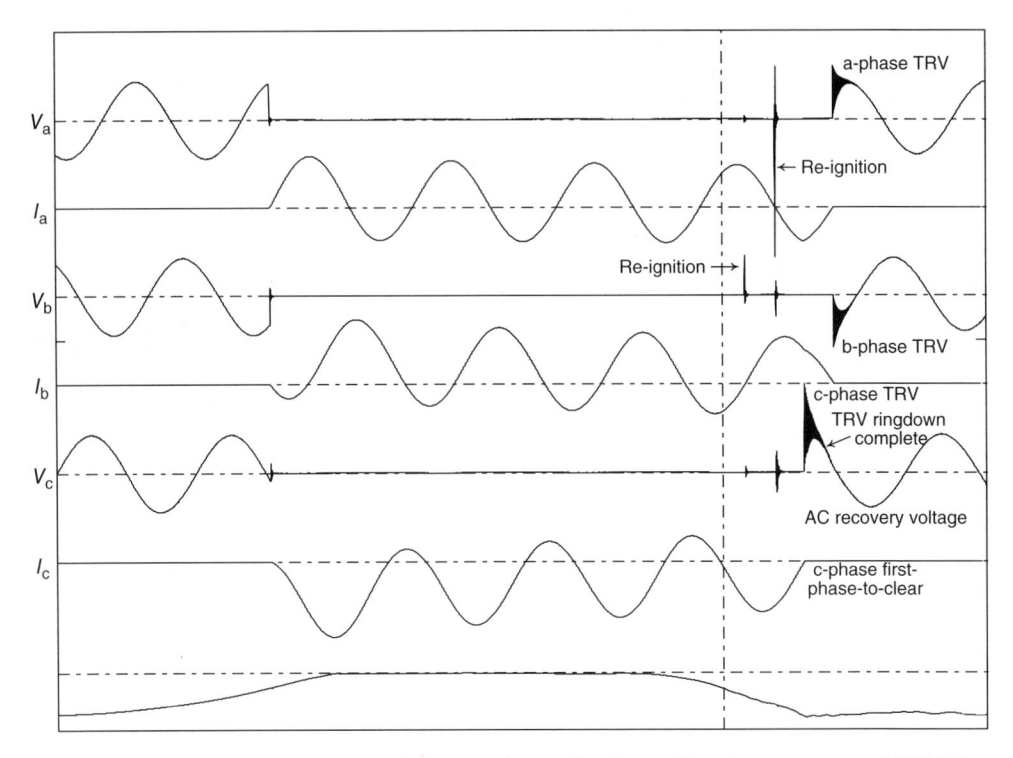

Figure 1.3 Current interruption terminology: voltage-related quantities (trace courtesy of KEMA).

- At current interruption in c-phase, the currents in a-phase and b-phase become equal in magnitude and opposite in polarity by means of a 30° shift, a shortened half-cycle in the former phase and a longer half-cycle in the latter. The total break time is the mechanical opening time plus the maximum arcing occurring in these two phases.

Figure 1.2 (current-related quantities):

- For a fault initiated at a voltage peak, the current will be symmetrical. Symmetrical means that the each half-cycle of the current, also referred to as a loop of current, will be identical to the preceding half-cycle of current. The current in a-phase is near symmetrical as a result of fault initiation just before the voltage peak.
- The currents in b-phase and c-phase are asymmetrical and consist of long and short loops of current referred to as major loops and minor loops, respectively. Maximum asymmetry occurs when the fault is initiated at a voltage zero crossing. Asymmetrical currents are discussed in detail in Section 4.7.

Figure 1.3 (voltage-related quantities):

- Current zeros occur every 60°, and the pole closest to a zero after contact part will make the first attempt to interrupt the current. The b-phase pole that is the closest to the first zero makes the attempt to interrupt the current but reignites because the contacts are too close to

withstand the TRV. The a-phase pole in turn also makes an attempt but reignites followed by successful interruption on c-phase, that is, recovering against the TRV and AC recovery voltage.
- The TRV is a transient oscillation as the voltage on the source side of the circuit breaker recovers to the prefault system voltage. The TRV oscillates around the AC recovery voltage, its aiming point or axis of oscillation, reaching a peak value depending on the damping in the circuit. As the trace shows, the TRV rings down within a power frequency quarter cycle. The first-pole-to-clear is exposed to the highest TRV. The theory behind TRVs is discussed in Chapters 2 and 3 and applied in the later chapters.
- a-phase and b-phase poles clear 90° later, each with its own TRV of lower magnitude than for c-phase and of opposite polarity. The AC recovery voltage is the line voltage and is shared by both poles.

Bibliography

The following references are for the textbooks covering the broad range of circuit breaker types and related switching transients in high-voltage networks (see also Bibliography in Appendix H).

1. Peterson, H.A. (1951) *Transients in Power Systems*, John Wiley & Sons, Inc.
2. Ragaller, K. (ed.) (1978) *Current Interruption in High-Voltage Networks*, Plenum Press.
3. Flurscheim, C.H. (ed.) (1982) *Power Circuit Breaker Theory and Practice*, Peter Peregrinus Ltd.
4. van der Sluis, L. (2001) *Transients in Power Systems*, John Wiley & Sons, Ltd.
5. Kapetanovich, M. (2011) *High Voltage Circuit Breakers*, Faculty of Electrotechnical Engineering (ETF), Sarajevo.
6. Janssen, A.L.J., Kapetanovich, M., Peelo, D.F., Smeets, R.P.P. and van der Sluis, L. (2014) *Switching in Electric Transmission and Distribution Systems*, John Wiley & Sons, Ltd.

2

RLC Circuits

2.1 General

The transients associated with transient recovery voltage (TRV), re-ignition or re-striking events can in general be related to either a series or parallel RLC circuit, each with specific boundary or initial conditions. The transients of most interest with respect to current interruption can, in fact, be represented by four basic RLC circuits, as shown in Figure 2.1. Each of these circuits will be treated in detail, and the following considerations apply to all cases:

1. All transients have a starting point, which may be zero or a finite value.
2. All transients have an axis of oscillation—or aiming point—that then becomes the ultimate steady-state value after the transient has died out.
3. All transients have a maximum value dependent on the degree of damping in the circuit.
4. All transients have a certain frequency determined by the values of L and C in the circuit; however, note that not all transients are oscillatory, the exceptions being the cases where the oscillations are aperiodic.

The calculation of the transient oscillations in each will be based on the general solutions of second-order linear homogeneous or non-homogeneous differential equations discussed in Appendix A. The boundary conditions to be used in each case are those of the value or rate of change of the value of the applicable current or voltage transient at time zero.

2.2 Series RLC Circuit with Step Voltage Injection

The circuit for this case is as shown in Figure 2.1a. The solutions for the transient current can be used to calculate inrush currents associated with single and back-to-back capacitor bank switching and currents associated with re-ignition and re-striking events.

Applying Kirchoff's voltage law to the circuit in Figure 2.1a, we can write

$$V = Ri(t) + L\frac{di(t)}{dt} + \frac{q(t)}{C}. \tag{2.1}$$

Current Interruption Transients Calculation, First Edition. David F. Peelo.
© 2014 John Wiley & Sons, Ltd. Published 2014 by John Wiley & Sons, Ltd.

RLC Circuits

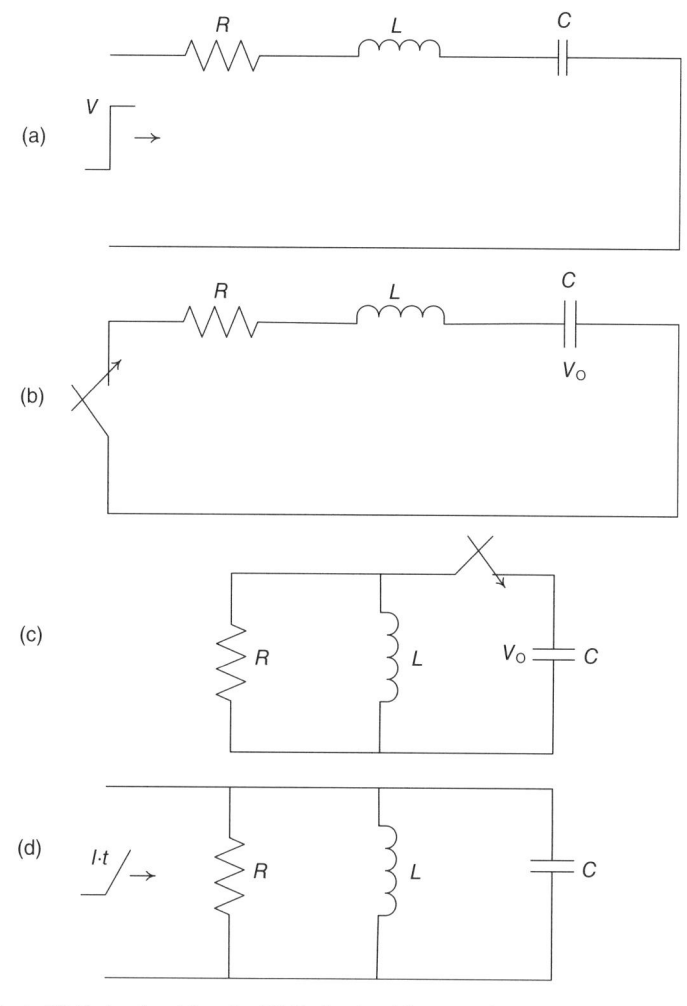

Figure 2.1 Basic RLC circuits: (a) series RLC circuit with step voltage injection; (b) source-free series RLC circuit with precharged capacitor; (c) source-free parallel RLC circuit with precharged capacitor; (d) parallel circuit with ramp current injection.

Differentiating Eq. (2.1) across then gives

$$L\frac{d^2i(t)}{dt^2} + R\frac{di(t)}{dt} + \frac{1}{C}i(t) = 0$$

or

$$\frac{d^2i(t)}{dt^2} + \frac{R}{L}\frac{di(t)}{dt} + \frac{1}{LC}i(t) = 0. \tag{2.2}$$

Equation (2.2) has the *abc* format discussed in Appendix A, and we can write

$$a=1, \quad b=\frac{R}{L}, \quad c=\frac{1}{LC}$$

and further that

$$\alpha=\frac{R}{2L} \quad \text{and} \quad \beta=\sqrt{\left(\frac{R}{2L}\right)^2-\frac{1}{LC}}=\sqrt{\alpha^2-\omega^2},$$

where

$$\omega=\frac{1}{\sqrt{LC}}.$$

Case 1: The circuit is overdamped ($\alpha^2 > \omega^2$).
From Table A.1, the solution for $i(t)$ is given by

$$i(t)=e^{-\alpha t}(k_1 \cosh \beta t + k_2 \sinh \beta t). \qquad (2.3)$$

To determine the values of k_1 and k_2, we must apply the boundary conditions given by $i(0)$ and $di(0)/dt$.
At $t=0$, $i(0)=0$ and $q=0$ and Eq. (2.3) becomes ($\sinh 0 = 0$ and $\cosh 0 = 1$)

$$0=e^{-0}(k_1(1)+k_2(0))$$

and therefore $k_1 = 0$, giving

$$i(t)=k_2 e^{-\alpha t} \sinh \beta t.$$

For the second boundary condition,

$$\frac{di(t)}{dt} = k_2 e^{-\alpha t}(\beta \cosh \beta t - \alpha \sinh \beta t)$$
$$= k_2 \beta \quad \text{at } t=0.$$

From Eq. (2.1), the current at time zero plus is determined by the inductance L, and we can write

$$V = R(0) + L\frac{di(0+)}{dt} + \frac{0}{C}$$

or

$$\frac{di(0+)}{dt} = \frac{V}{L}$$

RLC Circuits

and

$$k_2 = \frac{V}{L\beta}.$$

The solution for $i(t)$ is

$$i(t) = \frac{V}{L\sqrt{\alpha^2 - \omega^2}} e^{-\alpha t} \sinh \sqrt{\alpha^2 - \omega^2}\, t. \tag{2.4}$$

Case 2: The circuit is critically damped ($\alpha^2 = \omega^2$).
From Table A.1, the solution for $i(t)$ is

$$i(t) = (k_1 + k_2 t) e^{-\alpha t}.$$

At $t = 0$, $i(0) = 0$:

$$0 = (k_1 + 0)e^{-0},$$

giving $k_1 = 0$ and

$$i(t) = k_2 t\, e^{-\alpha t},$$
$$\frac{di(t)}{dt} = k_2\, e^{-\alpha t}(1 - \alpha t),$$
$$\frac{di(0+)}{dt} = k_2$$
$$= \frac{V}{L}.$$

The solution for $i(t)$ is

$$i(t) = \frac{V}{L} t\, e^{-\alpha t}. \tag{2.5}$$

Case 3: The circuit is underdamped ($\omega^2 > \alpha^2$).
From Table A.1, the solution for $i(t)$ is

$$i(t) = e^{-\alpha t}(k_1 \cos \beta t + k_2 \sin \beta t).$$

Applying the boundary conditions gives the same results as for Case 1:

$$k_1 = 0,$$
$$k_2 = \frac{V}{L\beta},$$

and the solution for $i(t)$ is

$$i(t) = \frac{V}{L\sqrt{\omega^2 - \alpha^2}} e^{-\alpha t} \sin \sqrt{\omega^2 - \alpha^2}\, t. \tag{2.6}$$

Equations (2.4)–(2.6) can be applied to any specific case for which R, L and C are known. However, it is of more interest to determine the behaviour of the circuit in general. This is done by expressing the damping in the circuit relative to critical damping and time relative to the period of the undamped oscillation, that is, as generic time.

To first consider circuit damping, the value of R to give critical damping (R_C) is determined as follows:

$$\alpha^2 = \omega^2$$

or

$$\left(\frac{R_C}{2L}\right)^2 = \frac{1}{LC},$$

giving

$$R_C = 2\sqrt{\frac{L}{C}}.$$

We now define the degree of damping d_s[1] in the circuit as

$$\begin{aligned}d_s &= \frac{\alpha}{\omega} \\ &= \frac{R}{2L} \frac{\sqrt{LC}}{1} \\ &= \frac{R}{2\sqrt{L/C}} \\ &= \frac{R}{R_C}.\end{aligned}$$

Substituting as applicable:

Overdamped case: $d_s > 1$, $R > R_C$.

$$\begin{aligned}\sqrt{\alpha^2 - \omega^2} &= \omega\sqrt{\left(\frac{\alpha}{\omega}\right)^2 - 1} \\ &= \omega\sqrt{d_s^2 - 1}.\end{aligned} \tag{2.7}$$

---

[1] d_s is sometimes referred to as a damping factor; however, this term tends to be used in another context (as discussed later) and therefore the term degree of damping is used throughout for RLC circuits.

RLC Circuits

Underdamped case: $d_s < 1$, $R < R_C$.

$$\sqrt{\omega^2 - \alpha^2} = \omega\sqrt{1 - \left(\frac{\alpha}{\omega}\right)^2} \quad (2.8)$$
$$= \omega\sqrt{1 - d_s^2}.$$

To derive generic time t_g, we can simply express real time t in terms of one period of the undamped oscillation $T = \sqrt{LC}$, giving

$$\begin{aligned} t_g &= \frac{t}{T} \\ &= \frac{t}{\sqrt{LC}} \\ &= \omega t, \end{aligned} \quad (2.9)$$

where t_g is in radians.

We can now substitute Eqs. (2.7) and (2.9) in Eq. (2.4), noting that $\alpha = \omega d_s$, giving

$$\begin{aligned} i(t_g) &= \frac{V}{L\omega\sqrt{d_s^2 - 1}} e^{-d_s \omega t} \sinh\sqrt{d_s^2 - 1}\,\omega t \\ &= \frac{V}{L \cdot (1/\sqrt{LC})\sqrt{d_s^2 - 1}} e^{-d_s t_g} \sinh\sqrt{d_s^2 - 1}\,t_g \\ &= \frac{V}{\sqrt{L/C}\sqrt{d_s^2 - 1}} e^{-d_s t_g} \sinh\sqrt{d_s^2 - 1}\,t_g \\ &= \frac{V}{Z\sqrt{d_s^2 - 1}} e^{-d_s t_g} \sinh\sqrt{d_s^2 - 1}\,t_g, \end{aligned} \quad (2.10)$$

where Z is the surge impedance of the circuit equal to $\sqrt{L/C}$ and V/Z is the value of the current if the circuit is undamped.

For the critically damped case, from Eqs. (2.5) and (2.9) ($d_s = 1$)

$$\begin{aligned} i(t_g) &= \frac{V t_g}{L \omega} e^{-d_s \omega t} \\ &= \frac{V}{L} \sqrt{LC}\, t_g\, e^{-t_g} \\ &= \frac{V}{\sqrt{L/C}}\, t_g\, e^{-t_g} \\ &= \frac{V}{Z}\, t_g\, e^{-t_g}. \end{aligned} \quad (2.11)$$

For the underdamped case, Eqs. (2.6), (2.8) and (2.9) apply, giving

$$i(t_g) = \frac{V}{Z\sqrt{1-d_s^2}} e^{-d_s t_g} \sin\sqrt{1-d_s^2}\, t_g. \qquad (2.12)$$

The last step is to express Eqs. (2.10)–(2.12) in pu by dividing across by V/Z, giving the following set of generic equations:

Overdamped case: $d_s > 1$, $R > R_C$.

$$i(t_g)_{pu} = \frac{e^{-d_s t_g}}{\sqrt{d_s^2-1}} \sinh\sqrt{d_s^2-1}\, t_g. \qquad (2.13)$$

Critically damped case: $d_s = 1$, $R = R_C$.

$$i(t)_{pu} = t_g\, e^{-t_g}. \qquad (2.14)$$

Underdamped case: $d_s < 1$, $R < R_C$.

$$i(t_g)_{pu} = \frac{e^{-d_s t_g}}{\sqrt{1-d_s^2}} \sin\sqrt{1-d_s^2}\, t_g. \qquad (2.15)$$

Generic damping curves based on Eqs. (2.13)–(2.15) may now be plotted as shown in Figure 2.2.

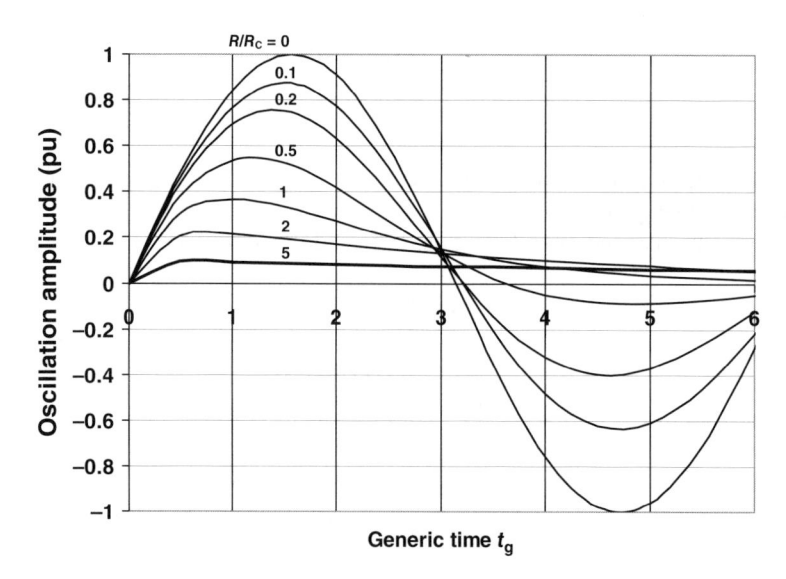

Figure 2.2 Generic damping curves for a series RLC circuit with step voltage injection.

RLC Circuits

The starting point of the transient oscillation is zero, and the axis of oscillation is also zero because the current must go to zero when the capacitor C is charged to V. Because t_g is in radians, one complete cycle of the undamped oscillation takes 2π radians.

2.3 Source-Free Series RLC Circuit with Precharged Capacitor

From Figure 2.1b, the circuit for this case is reproduced as Figure 2.3. The solutions for the transient voltage at the capacitor can be used to calculate transient recovery voltages for inductive load switching and voltage transients associated with re-ignition and re-striking events.

Applying Kirchoff's current law to the current in Figure 2.3, we can write

$$v_c(t) + Ri(t) + L\frac{di(t)}{dt} = 0, \tag{2.16}$$

where $v_c(t)$ is the voltage at the capacitor.

We can also write

$$i(t) = \frac{dq(t)}{dt} = C\frac{dv_c(t)}{dt}$$

and substituting in Eq. (2.16) gives

$$\frac{d^2 v_c(t)}{dt^2} + \frac{R}{L}\frac{dv_c(t)}{dt} + \frac{1}{LC}v_c(t) = 0.$$

From Appendix A,

$$a = 1, \quad b = \frac{R}{L}, \quad c = \frac{1}{LC},$$

$$\alpha = \frac{R}{2L},$$

$$\beta = \sqrt{\left(\frac{R}{2L}\right)^2 - \frac{1}{LC}}$$

$$= \sqrt{\alpha^2 - \omega^2}.$$

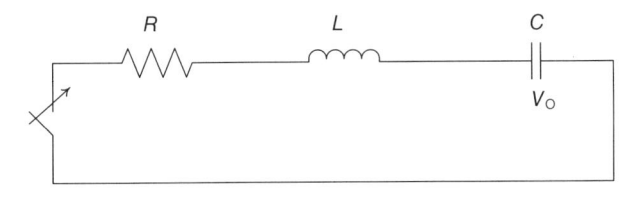

Figure 2.3 Source-free series RLC circuit with precharged capacitor.

The boundary conditions at time zero are

$$v_c(0) = V_0 \quad \text{(the initial charge on the capacitor)},$$
$$\frac{dv_c(0)}{dt} = 0 \quad \text{(due to the inductance } L \text{ in the circuit).}$$

Case 1: The circuit is overdamped ($\alpha^2 > \omega^2$).
From Table A.1, the solution for $v_c(t)$ is

$$v_c(t) = e^{-\alpha t}(k_1 \cosh \beta t + k_2 \sinh \beta t).$$

Applying the boundary conditions in turn,

$$V_0 = e^0(k_1(1) + k_2(0)),$$
$$k_1 = V_0,$$
$$\frac{dv_c}{dt} = V_0 e^{-\alpha t}(-\alpha \cosh \beta t + \beta \sinh \beta t) + k_2 e^{-\alpha t}(-\alpha \sinh \beta t + \beta \cosh \beta t)$$

and

$$\frac{dv_c(0)}{dt} = V_0 e^{-0}(-\alpha(1) + \beta(0)) + k_2 e^{-0}(-\alpha(0) + \beta(1))$$
$$= -\alpha V_0 + \beta k_2$$
$$= 0,$$
$$k_2 = \frac{\alpha}{\beta} V_0.$$

The solution for $v_c(t)$ is

$$v_c(t) = V_0 e^{-\alpha t}\left(\cosh \beta t + \frac{\alpha}{\beta} \sinh \beta t\right). \tag{2.17}$$

Case 2: The circuit is critically damped ($\alpha^2 = \omega^2$).
From Table A.1, the solution for $v_c(t)$ is

$$v_c(t) = (k_1 + k_2 t)e^{-\alpha t}.$$

Applying the boundary conditions at time zero,

$$V_0 = (k_1 + k_2(0))e^{-0},$$
$$k_1 = V_0,$$
$$\frac{dv_c(t)}{dt} = -\alpha V_0 e^{-\alpha t} + k_2 e^{-\alpha t}(-\alpha t + 1),$$
$$\frac{dv_c(0)}{dt} = -\alpha V_0 e^{-0} + k_2 e^{-0}(-\alpha(0) + 1)$$
$$= 0,$$
$$k_2 = \alpha V_0.$$

RLC Circuits

The solution for $v_c(t)$ is

$$v_c(t) = V_0 \, e^{-\alpha t}(1 + \alpha t). \tag{2.18}$$

Case 3: The circuit is underdamped ($\omega^2 > \alpha^2$).
From Table A.1, the solution for $v_c(t)$ is

$$v_c(t) = e^{-\alpha t}(k_1 \cos \beta t + k_2 \sin \beta t).$$

Applying the boundary conditions at time zero gives a result similar to that for Case 1:

$$k_1 = V_0,$$
$$k_2 = \frac{\alpha}{\beta} V_0,$$

and the solution for $v_c(t)$ is

$$v_c(t) = V_0 \, e^{-\alpha t}\left(\cos \beta t + \frac{\alpha}{\beta} \sin \beta t\right). \tag{2.19}$$

As earlier, Eqs. (2.17)–(2.19) can be converted to per-unit generic equations (dividing across by V_0) by making the following substitutions:

$$d_s = \frac{\alpha}{\omega} = \frac{R}{R_C},$$
$$t_g = \omega t,$$
$$\alpha t \to d_s \omega t = d_s t_g$$
$$\beta = \sqrt{\alpha^2 - \omega^2}$$
$$= \omega \sqrt{d_s^2 - 1} \quad \text{for Case 1,}$$
$$\beta = \omega \sqrt{1 - d_s^2} \quad \text{for Case 3,}$$
$$\frac{\alpha}{\beta} = \frac{\alpha}{\omega \sqrt{d_s^2 - 1}}$$
$$= \frac{d_s}{\sqrt{d_s^2 - 1}} \quad \text{for Case 1,}$$
$$\frac{\alpha}{\beta} = \frac{d_s}{\sqrt{1 - d_s^2}} \quad \text{for Case 3.}$$

The per-unit generic equations are as follows:

Case 1: The circuit is overdamped ($d_s > 1$, $R > R_C$).

$$v_c(t)_{pu} = e^{-d_s t_g}\left(\cosh \sqrt{d_s^2 - 1}\, t_g + \frac{d_s}{\sqrt{d_s^2 - 1}} \sinh \sqrt{d_s^2 - 1}\, t_g\right). \tag{2.20}$$

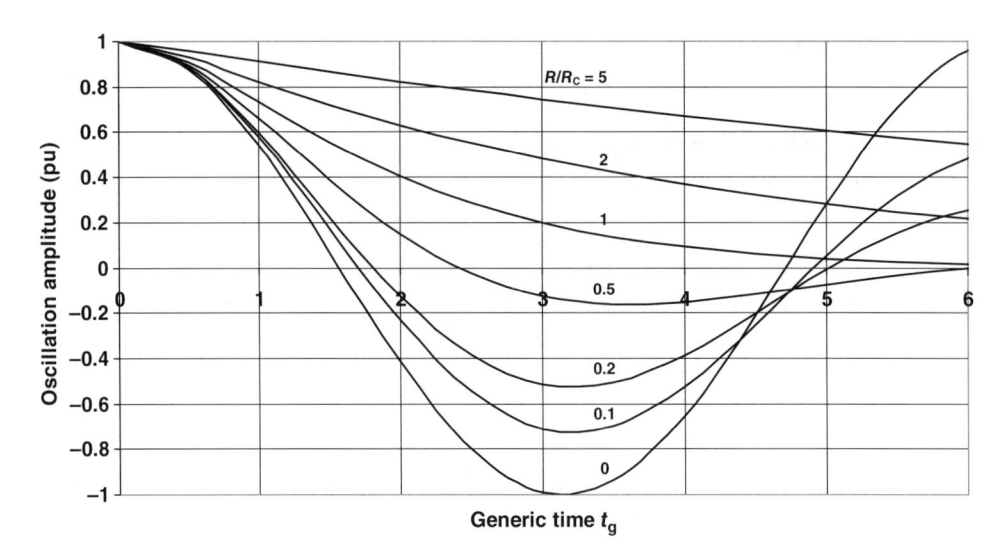

Figure 2.4 Generic damping curves for series RLC circuit with precharged capacitor.

Case 2: The circuit is critically damped ($d_s = 1$, $R = R_C$).

$$v_c(t)_{pu} = e^{-t_g}(1 + t_g). \tag{2.21}$$

Case 3: The circuit is underdamped ($d_s < 1$, $R < R_C$).

$$v_c(t)_{pu} = e^{-d_s t_g}\left(\cos\sqrt{1 - d_s^2}\, t_g + \frac{d_s}{\sqrt{1 - d_s^2}}\sin\sqrt{1 - d_s^2}\, t_g\right). \tag{2.22}$$

Generic damping curves based on Eqs. (2.20)–(2.22) are shown in Figure 2.4. The starting point of the oscillations is at a defined voltage and the axis of oscillation is zero.

2.4 Source-Free Parallel RLC Circuit with Precharged Capacitor

From Figure 2.1c, the circuit for this case is reproduced in Figure 2.5. This circuit and case is a variation of the series case discussed in Section 2.3 and the solutions can be used for the same purposes.

Applying Kirchoff's current law, we can write

$$\frac{v(t)}{R} + \frac{1}{L}\int v(t)dt + C\frac{dv(t)}{dt} = 0, \tag{2.23}$$

where $v(t)$ is the node voltage common to all three elements. Differentiating across Eq. (2.23)

RLC Circuits

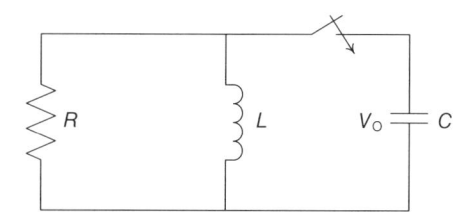

Figure 2.5 Source-free parallel RLC circuit with precharged capacitor.

takes on the *abc* format

$$\frac{d^2v(t)}{dt^2} + \frac{1}{RC}\frac{dv(t)}{dt} + \frac{1}{LC}v(t) = 0.$$

The boundary conditions at time zero are

$$v(0) = V_0$$

and

$$\frac{dv(0)}{dt} = -\frac{V_0}{RC},$$

which is determined from Eq. (2.23), recognizing that there is initial current in R only.

The calculation of the overdamped, critically damped and underdamped equations follows the same process used in Sections 2.2 and 2.3. It is left to readers to show that the solutions are as follows:

Overdamped case:

$$v(t) = V_0 \, e^{-\alpha t}\left(\cosh \beta t - \frac{\alpha}{\beta}\sinh \beta t\right). \tag{2.24}$$

Critically damped case:

$$v(t) = V_0 \, e^{-\alpha t}(1 - \alpha t). \tag{2.25}$$

Underdamped case:

$$v(t) = V_0 \, e^{-\alpha t}\left(\cos \beta t - \frac{\alpha}{\beta}\sin \beta t\right). \tag{2.26}$$

To derive the generic equations, we define the degree of damping, now designated as d_p, as

$$d_p = \frac{\alpha}{\omega}, \tag{2.27}$$

where
$$\alpha = \frac{1}{2RC} \quad \text{and} \quad \omega = \frac{1}{\sqrt{LC}}.$$

For critical damping $R = R_C$,
$$\frac{1}{2R_C C} = \frac{1}{LC},$$

giving
$$R_C = 0.5\sqrt{\frac{L}{C}}.$$

From Eq. (2.27),
$$d_p = \frac{1}{2RC}\sqrt{\frac{LC}{1}}$$
$$= \frac{1}{R}\left(0.5\sqrt{\frac{L}{C}}\right)$$
$$= \frac{R_C}{R}.$$

d_p is the opposite of d_s. This is logical because damping in a series circuit decreases with decreasing R, but in a parallel circuit, damping decreases with increasing R. For the totally undamped events, that is, $R = 0$, the series and parallel circuits are the same and described by the same equation. d_p and d_s, however, are not inversely related because R_C is not the same in each case; that is, a general duality does not exist (Appendix C).

The per-unit generic equations are as follows:

Case 1: The circuit is overdamped ($d_p > 1$, $R < R_C$).

$$v(t)_{pu} = e^{-d_p t_g}\left(\cosh\sqrt{d_p^2 - 1}\, t_g - \frac{d_p}{\sqrt{d_p^2 - 1}}\sinh\sqrt{d_p^2 - 1}\, t_g\right). \qquad (2.28)$$

Case 2: The circuit is critically damped ($R = R_C$).

$$v(t)_{pu} = e^{-t_g}(1 - t_g). \qquad (2.29)$$

Case 3: The circuit is underdamped ($d_p < 1$, $R > R_C$).

$$v(t)_{pu} = e^{-d_p t_g}\left(\cos\sqrt{1 - d_p^2}\, t_g - \frac{d_p}{\sqrt{1 - d_p^2}}\sin\sqrt{1 - d_p^2}\, t_g\right). \qquad (2.30)$$

RLC Circuits

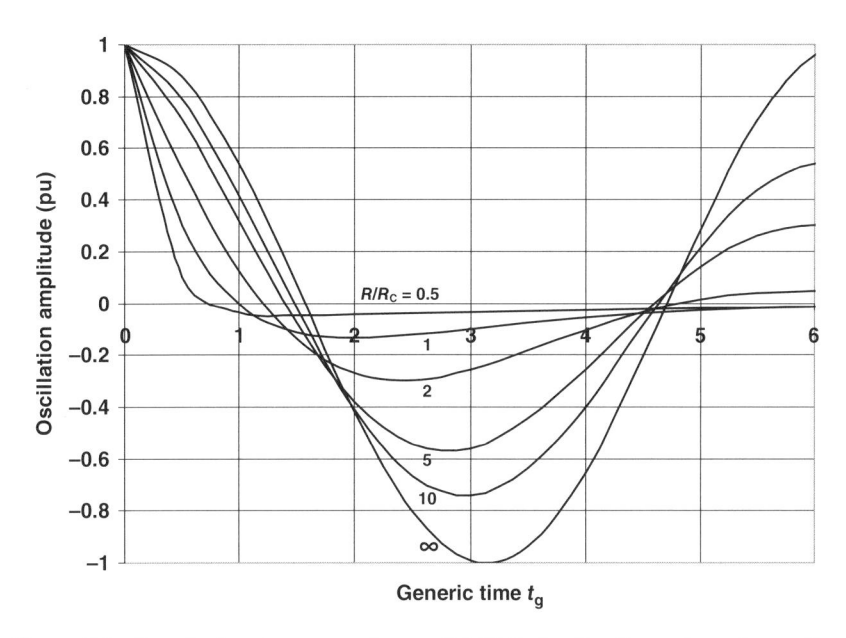

Figure 2.6 Generic damping curves for parallel RLC circuit with precharged capacitor.

Generic damping curves based on Eqs. (2.28)–(2.30) are shown in Figure 2.6.

As for the series case, the starting point of the oscillations is at a defined voltage and the axis of oscillation is zero.

2.5 Parallel RLC Circuit with Ramp Current Injection

From Figure 2.1d, the circuit for this case is reproduced in Figure 2.7. The solutions for this case are used to calculate circuit breaker TRVs and variations to them caused by the addition of surge or shunt capacitors or opening resistors.

The notion of current injection to cancel fault current and calculate the circuit voltage response was first published by Boehne [2] but Bewley [4] suggests that the notion originates from Heaviside. Referring to Figure 2.7, the intent is to calculate the voltage $v(t)$ on the common node because of the injected ramp current I A/s.

Applying Kirchoff's current law, we can write

$$\frac{v(t)}{R} + \frac{1}{L}\int v(t)dt + C\frac{dv(t)}{dt} = I \cdot t,$$

which can in turn be rewritten as

$$\frac{d^2v(t)}{dt^2} + \frac{1}{RC}\frac{dv(t)}{dt} + \frac{1}{LC}v(t) = \frac{I}{C}. \tag{2.31}$$

As discussed in Appendix A, Eq. (2.31) is a second-order non-homogeneous differential equation. The right-hand side is a constant, and from Eq. (A.22), the solution is

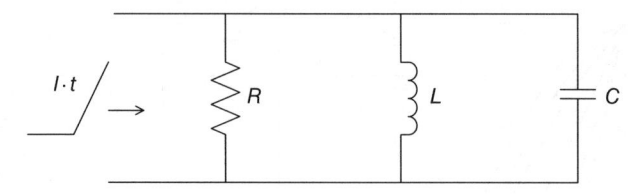

Figure 2.7 Parallel RLC circuit with ramp current injection.

$$v(t) = k_1 e^{r_1 t} + k_2 e^{r_2 t} + \frac{K}{c},$$

where K is the constant I/C and c is $1/LC$.

The solution for Eq. (2.31) is then

$$v(t) = k_1 e^{r_1 t} + k_2 e^{r_2 t} + IL \qquad (2.32)$$

noting that the dimension of IL is volts and reflects the voltage drop across the source impedance during a fault as discussed further in Chapter 4.

At time zero, the boundary conditions are

$$v(0) = 0 \quad \text{and} \quad \frac{dv(0)}{dt} = 0,$$

and we can solve for the overdamped, critically damped and underdamped cases using the generic solutions in Table A.1.

Case 1: The circuit is overdamped ($\alpha^2 > \omega^2$).

From Table A.1, Eq. (2.32) can be rewritten as

$$v(t) = e^{-\alpha t}(k_1 \cosh \beta t + k_2 \sinh \beta t) + LI. \qquad (2.33)$$

Applying the boundary conditions at $t = 0$,

$$v(0) = (1)(k_1(1) + k_2(0)) + LI$$
$$= 0,$$
$$k_1 = -LI,$$
$$\frac{dv(t)}{dt} = -\alpha e^{-\alpha t}(k_1 \cosh \beta t + k_2 \sinh \beta t) + e^{-\alpha t}(\beta k_1 \sinh \beta t + \beta k_2 \cosh \beta t),$$
$$\frac{dv(0)}{dt} = -\alpha(1)(k_1(1) + k_2(0)) + (1)(\beta k_1(0) + \beta k_2(1))$$
$$= 0,$$
$$k_2 = \frac{\alpha}{\beta} k_1$$
$$= -LI \frac{\alpha}{\beta}.$$

RLC Circuits

Substituting for k_1 and k_2 in Eq. (2.33),

$$v(t) = LI\left[1 - e^{-\alpha t}\left(\cosh \beta t + \frac{\alpha}{\beta}\sinh \beta t\right)\right]. \tag{2.34}$$

Expressed in per unit (dividing across by LI), Eq. (2.34) becomes

$$v(t)_{\text{pu}} = 1 - e^{-\alpha t}\left(\cosh \beta t + \frac{\alpha}{\beta}\sinh \beta t\right). \tag{2.35}$$

Case 2: The circuit is critically damped ($\alpha^2 = \omega^2$).
From Table A.1, Eq. (2.32) can be written as

$$v(t) = (k_1 + k_2 t)e^{-\alpha t} + LI. \tag{2.36}$$

It is left to readers to show that applying the boundary conditions to Eq. (2.36) will give the following solutions:

$$v(t) = LI\left[1 - F^{-\alpha t}(1 + \alpha t)\right]$$

and

$$v(t)_{\text{pu}} = 1 - e^{-\alpha t}(1 + \alpha t). \tag{2.37}$$

Case 3: The circuit is underdamped ($\alpha^2 < \omega^2$).
From Table A.1, Eq. (2.32) can be written as

$$v(t) = e^{-\alpha t}(k_1 \cos \beta t + k_2 \sin \beta t) + LI. \tag{2.38}$$

It is again left to readers to show that applying the boundary conditions to Eq. (2.38) will give the following solutions:

$$v(t) = LI\left[1 - e^{-\alpha t}\left(\cos \beta t + \frac{\alpha}{\beta}\sin \beta t\right)\right]$$

and

$$v(t)_{\text{pu}} = 1 - e^{-\alpha t}\left(\cos \beta t + \frac{\alpha}{\beta}\sin \beta t\right). \tag{2.39}$$

To derive the generic equations, we can again apply the degree of damping d_{p} as for the parallel RLC circuit discussed in Section 2.4.

The per-unit generic equations follow from Eqs. (2.35), (2.37) and (2.39):

Case 1: The circuit is overdamped ($d_p > 1$, $R < R_C$).

$$v(t)_{pu} = 1 - e^{-d_p t_g}\left(\cosh\sqrt{d_p^2 - 1}\, t_g + \frac{d_p}{\sqrt{d_p^2 - 1}}\sinh\sqrt{d_p^2 - 1}\, t_g\right). \quad (2.40)$$

Case 2: The circuit is critically damped ($R = R_C$).

$$v(t)_{pu} = 1 - e^{-t_g}\left(1 + t_g\right). \quad (2.41)$$

Case 3: The circuit is underdamped ($d_p < 1$, $R > R_C$).

$$v(t)_{pu} = 1 - e^{-d_p t_g}\left(\cos\sqrt{1 - d_p^2}\, t_g + \frac{d_p}{\sqrt{1 - d_p^2}}\sin\sqrt{1 - d_p^2}\, t_g\right). \quad (2.42)$$

Generic damping curves based on Eqs. (2.40)–(2.42) are shown in Figure 2.8.

As noted earlier, transients can be described as having a starting point, an aiming point or axis of oscillation and a peak value. Referring to Figure 2.4, for example, the starting point is 1 pu, the axis of oscillation is zero and the peak value depends on the degree of damping. For the

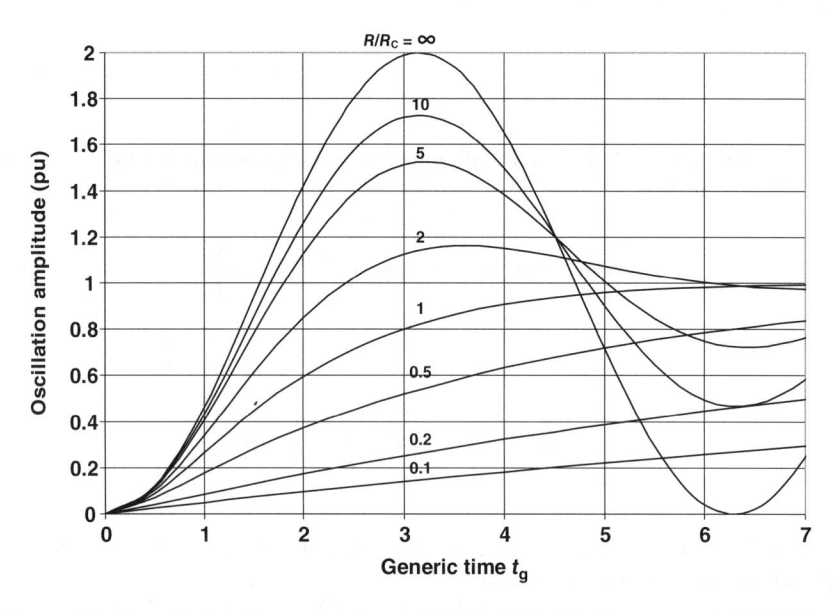

Figure 2.8 Generic damping curves for parallel RLC circuit with ramp current injection.

case in Figure 2.8, the starting point is zero, the axis of oscillation is 1 pu and again the peak value depends on the degree of damping. To apply the various curves to actual switching cases, values will be assigned to these quantities and to t_g.

For underdamped oscillations, the peak value is described by the so-called amplitude factor k_{af}. This factor is the ratio of the peak value to the axis of oscillation both relative to the starting point. The amplitude factors for the cases shown in Figures 2.4 and 2.8 are illustrated in Figures 2.9 and 2.10, respectively. The maximum possible value for k_{af} is 2 pu.

As noted earlier, the term damping factor is also used in this context. Referring to either Figure 2.9 or Figure 2.10, the damping factor β (usually designated by this Greek letter and not to be confused with the β in the transient calculations) is given by

$$\beta = \frac{B-A}{A}.$$

On this basis,

$$k_{af} = 1 + \beta.$$

The maximum possible value that β can assume is 1.

A further relationship of interest is that between the amplitude factor k_{af} and the degree of damping for the underdamped curves in Figure 2.8. From Eq. (2.42), we can write

$$k_{af} = 1 - e^{-d_p t_g}\left(\cos\sqrt{1-d_p^2}\,t_g + \frac{d_p}{\sqrt{1-d_p^2}}\sin\sqrt{1-d_p^2}\,t_g\right). \qquad (2.43)$$

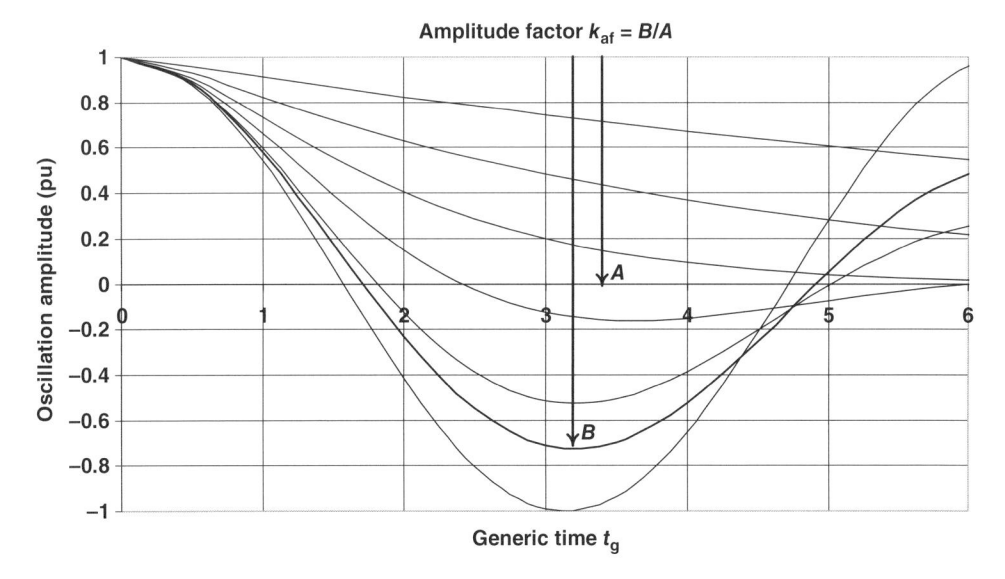

Figure 2.9 Amplitude factor for oscillation starting at 1 pu (see Figure 2.4).

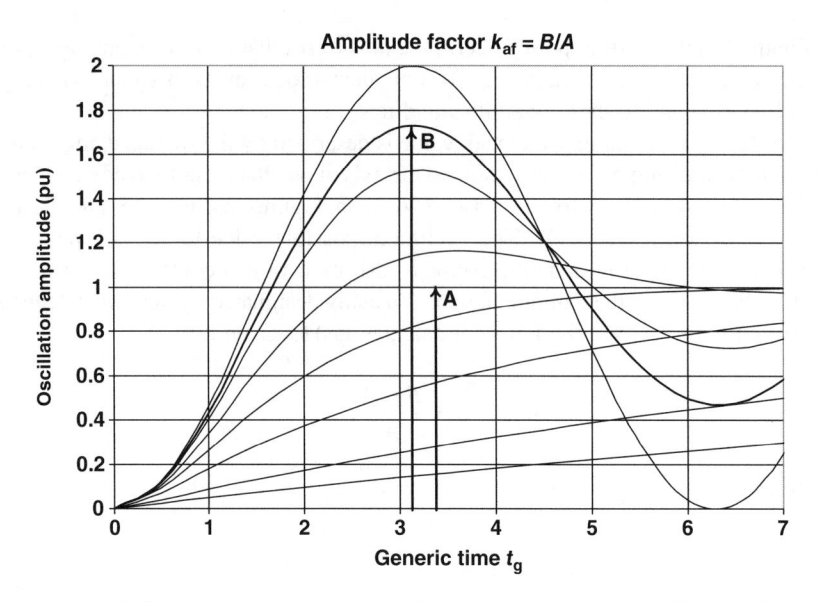

Figure 2.10 Amplitude factor for oscillation starting at zero (see Figure 2.8).

Differentiating Eq. (2.43) and setting the result to zero shows that the peak value occurs at a time

$$t_g = \frac{\pi}{\sqrt{1-d_p^2}}$$

and substitution in Eq. (2.43) gives

$$k_{af} = 1 - e^{-\pi d_p/\sqrt{1-d_p^2}} \left(\cos \pi + \frac{d_p}{\sqrt{1-d_p^2}} \sin \pi \right) \qquad (2.44)$$

$$= 1 + e^{-\pi d_p/\sqrt{1-d_p^2}}.$$

A more user-friendly equation is obtained by expressing Eq. (2.44) in terms of R/R_C:

$$k_{af} = 1 + e^{-A}, \qquad (2.45)$$

where $A = \pi/\sqrt{(R/R_C)^2 - 1}$.

Equation (2.45) is plotted in Figure 2.11.

… RLC Circuits

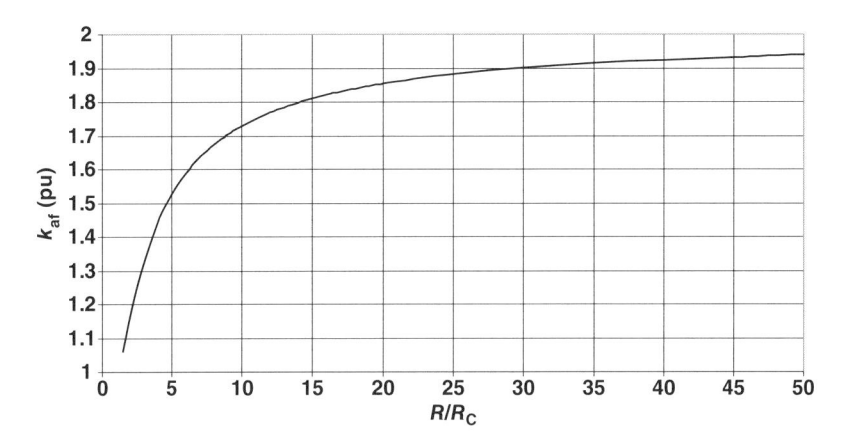

Figure 2.11 Amplitude factor k_{af} as a function of degree of damping for the underdamped curves of Figure 2.8.

The reverse equation, that is, k_{af} known and to calculate the value of R/R_C, is

$$\frac{R}{R_C} = \sqrt{1 + \left(\frac{\pi}{\ln(1/(k_{af}-1))}\right)^2}, \tag{2.46}$$

which readers can readily verify.

2.6 Alternative Equations

The credit for using a generic approach to transient analysis is due to Lackey who presented the curves of Figure 2.8 in his 1951 paper [3]. Later, Greenwood [5] expanded the approach to the other transient cases but chose to use a degree of damping (η) defined as

$$\eta = \frac{R}{\sqrt{L/C}} = \frac{R}{Z}$$

rather than using Lackey's R/R_C quantity. The relationship between d_p and η for the case in Section 2.5 is

$$d_p = \frac{1}{2\eta}. \tag{2.47}$$

Equations (2.40)–(2.42) can be converted to incorporate η and, taking the latter equation as an example, we can write

$$\frac{d_p}{\sqrt{1-d_p^2}} \rightarrow \frac{1}{\sqrt{4\eta^2-1}},$$

$$e^{-d_p t_g} \rightarrow e^{-t_g/2\eta},$$

$$\sqrt{1-d_p^2}\, t_g \rightarrow \sqrt{4\eta^2-1}\, \frac{t_g}{2\eta},$$

giving

$$v(t)_{pu} = 1 - e^{-t_g/2\eta}\left(\cos\sqrt{4\eta^2-1}\,t_g/2\eta + \frac{1}{\sqrt{4\eta^2-1}}\sin\sqrt{4\eta^2-1}\,t_g/2\eta\right). \quad (2.48)$$

Equation (2.48) will, of course, give the same result as Eq. (2.42), but its format is clearly the more convoluted. From Eq. (2.47), critical damping occurs for $\eta = 0.5$.

Greenwood also introduced the notion of duality between series and parallel RLC circuits; this is discussed further in the following chapters and in Appendix B. Clearly, the solutions of the series and parallel circuits have the same basic equations (Appendix A) but readers should satisfy themselves that a true duality exists before applying the principle.

2.7 Travelling Wave Basics

Travelling waves in power systems are mostly associated with lightning strikes and switching surges caused by circuit closing events such as high-speed reclosing after a successful fault clearing. However, in certain fault clearing cases, the TRVs generate or are actually travelling waves. Basic considerations are therefore provided here for later use in Chapters 4 and 5.

Travelling waves follow the laws of electricity, and Ohm's law and Kirchoff's voltage and current laws are applicable. Taking the representation shown in Figure 2.12, an incident voltage wave v_i travels on a transmission line and is accompanied by an incident current wave i_i. A reflection occurs at the line termination (discontinuity), generating a reflected voltage wave v_r and a reflected current wave $-i_r$, the minus sign indicating the reversed direction of the current. The ratio of the respective voltages and currents gives the surge impedance Z of the line, and we can write

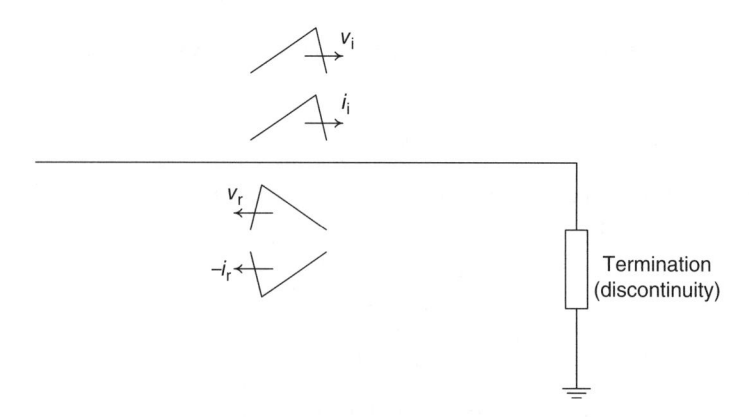

$$\frac{v_i}{i_i} = \frac{v_r}{-i_r} = Z$$

Figure 2.12 Travelling waves on a terminated transmission line.

RLC Circuits

and

$$Z = \sqrt{\frac{L}{C}},$$

where L and C are the per-unit length inductance and capacitance of the transmission line, respectively. Note that Z is resistive, and the above incident and reflected voltages and currents are in phase.

For any time and location, the voltage v and the current i are given by

$$v = v_i + v_r \tag{2.49}$$

and

$$i = i_i + i_r = \frac{v_i}{Z} - \frac{v_r}{Z}$$

or

$$Zi = v_i - v_r. \tag{2.50}$$

From Eqs. (2.49) and (2.50), it follows that

$$Zi + v = 2v_i. \tag{2.51}$$

Equation (2.51) is the fundamental equation for dealing with the effect of transmission line terminations. The cases of interest with respect to TRVs are line open-circuited, line short-circuited and line terminated by an inductance L_t.

Case 1: Line open-circuited.

$$i = 0,$$
$$v = 2v_i.$$

This is voltage doubling, meaning that the reflected wave is equal to and adds to the incident wave.

Case 2: Line short-circuited.

$$v = 0,$$
$$i = \frac{2v_i}{Z}.$$

This is current doubling, and because $v = 0$, a reflected voltage wave of opposite polarity to the incident wave is generated to cancel the latter wave.

Table 2.1 Series RLC circuit equations.

Case	Series RLC	
	Current; no prior charge on capacitor	Voltage; source–free capacitor precharged
Circuit	(R, L, C series with source V, current i)	(R, L, C series, capacitor precharged to V_0)
Boundary conditions	$i(0)=0$, $\dfrac{di(0+)}{dt}=V/L$	$v(0)=V_0$, $\dfrac{dv(0)}{dt}=0$
Basic equation	$\dfrac{d^2 i}{dt^2}+\dfrac{R}{L}\dfrac{di}{dt}+\dfrac{1}{LC}i=0$	$\dfrac{d^2 v}{dt^2}+\dfrac{R}{L}\dfrac{dv}{dt}+\dfrac{1}{LC}v=0$
Degree of damping	$d_s=\dfrac{R}{R_C}$, $R_C=2\sqrt{\dfrac{L}{C}}=2Z$	$d_s=\dfrac{R}{R_C}$, $R_C=2\sqrt{\dfrac{L}{C}}=2Z$
Overdamped	$i(t)=\dfrac{V}{Z\sqrt{d_s^2-1}}\,e^{-d_s t_g}\sinh\sqrt{d_s^2-1}\,t_g$	$v(t)=e^{-d_s t_g}\left[\cosh\sqrt{d_s^2-1}\,t_g+\dfrac{d_s}{\sqrt{d_s^2-1}}\sinh\sqrt{d_s^2-1}\,t_g\right]$
Critically damped	$i(t)=\dfrac{V}{Z}t_g\,e^{-t_g}$	$v(t)=e^{-t_g}(1+t_g)$
Underdamped	$i(t)=\dfrac{V}{Z\sqrt{1-d_s^2}}\,e^{-d_s t_g}\sin\sqrt{1-d_s^2}\,t_g$	$v(t)=e^{-d_s t_g}\left[\cos\sqrt{1-d_s^2}\,t_g+\dfrac{d_s}{\sqrt{1-d_s^2}}\sin\sqrt{1-d_s^2}\,t_g\right]$
Use for	Inrush single + back-to-back capacitor banks; re-strike + re-ignition currents for reactive load circuits	TRVs for inductive load circuits; re-striking inductive + capacitive load circuits

Note: t_g is generic time given by $t_g = t/\sqrt{LC}$, where t is real time.

RLC Circuits

Table 2.2 Parallel RLC circuit equations.

Case	Parallel RLC	Parallel RLC
	Voltage; source–free capacitor precharged	Voltage; ramp current injection
Circuit	(R, L, C with switch, V_0)	(I·t source, R, L, C)
Boundary conditions	$v(0) = V_0, \quad \dfrac{dv(0)}{dt} = -\dfrac{V_0}{RC}$	$v(0) = 0, \quad \dfrac{dv(0)}{dt} = 0$
Basic equation	$\dfrac{d^2v}{dt^2} + \dfrac{1}{RC}\dfrac{dv}{dt} + \dfrac{1}{LC}v = 0$	$\dfrac{d^2v}{dt^2} + \dfrac{1}{RC}\dfrac{dv}{dt} + \dfrac{1}{LC}v = \dfrac{I}{C}$
Degree of damping	$d_p = \dfrac{R_C}{R}, \quad R_C = 0.5\sqrt{\dfrac{L}{C}} = 0.5Z$	$d_p = \dfrac{R_C}{R}, \quad R_C = 0.5\sqrt{\dfrac{L}{C}} = 0.5Z$
Overdamped	$v(t) = e^{-d_p t_g}\left[\cosh\sqrt{d_p^2-1}\,t_g - \dfrac{d_p}{\sqrt{d_p^2-1}}\sinh\sqrt{d_p^2-1}\,t_g\right]$	$v(t) = \left[1 - e^{-d_p t_g}\left(\cosh\sqrt{d_p^2-1}\,t_g + \dfrac{d_p}{\sqrt{d_p^2-1}}\sinh\sqrt{d_p^2-1}\,t_g\right)\right]$
Critically damped	$v(t) = e^{-t_g}(1 - t_g)$	$v(t) = 1 - e^{-t_g}(1 + t_g)$
Underdamped	$v(t) = e^{-d_p t_g}\left[\cos\sqrt{1-d_p^2}\,t_g - \dfrac{d_p}{\sqrt{1-d_p^2}}\sin\sqrt{1-d_p^2}\,t_g\right]$	$v(t) = \left[1 - e^{-d_p t_g}\left(\cos\sqrt{1-d_p^2}\,t_g + \dfrac{d_p}{\sqrt{1-d_p^2}}\sin\sqrt{1-d_p^2}\,t_g\right)\right]$
Use for	Variation of series precharged capacitor case	TRVs for terminal faults

Note: t_g is generic time given by $t_g = t/\sqrt{LC}$, where t is real time.

Case 3: Line terminated by an inductance L_t.

$$v = L_t \frac{di}{dt},$$

$$Zi + L_t \frac{di}{dt} = 2v_i. \qquad (2.52)$$

For Eq. (2.52), the solution is dependent on the form of v_i, that is, step function, ramp function, exponential function and so on.

2.8 Summary

In this chapter, we have shown that RLC circuit transients can be treated on a general basis. Starting with the common second-order linear differential equations (Appendix A) and then applying well-known trigonometric and exponential functions (as provided in Appendix C), sets of equations are derived. The equations and their application cases are summarized in Tables 2.1 and 2.2.

In later chapters, we will use the equations for specific cases of fault clearing and inductive load current interruption TRVs and inrush and re-striking currents. The calculations, as we have already seen, require a knowledge of the boundary conditions and generally these are given by considering the quantity in question—voltage or current—and its rate of change at time zero. This will yield two equations, enabling the derivation of two constants.

Bibliography

1. Shenkman, A. (2005) *Transient Analysis of Electric Power Circuits Handbook*, Springer.
2. Boehne, E.W. (1935) The determination of circuit recovery rates. *AIEE Trans.*, **54**, 530–539.
3. Lackey, C.H.W. (1951) Some technical considerations relating to design, performance and application of high-voltage switchgear. *Trans. S. Afr. Inst. Electr. Eng.*, 261–298.
4. Bewley, L.W. (1939) Travelling waves initiated by switching. *AIEE Trans.*, **58**, 18–26.
5. Greenwood, A. (1971) *Electrical Transients in Power Systems*, John Wiley & Sons, Inc.

3

Pole Factor Calculation

3.1 General

The analysis of balanced three-phase AC systems can generally be done using a single-phase approach. For unbalanced systems, the single-phase approach cannot be used, and analysis is that much more difficult. However, in 1918, Fortescue published his classic paper showing that a three-phase unbalanced system could be transformed into a set of balanced components known as symmetrical components.[1]

The symmetrical component set is shown in Figure 3.1 and consists of positive-, negative- and zero-sequence components, the sum of which equals exactly the original unbalanced system. The positive-sequence components have a phase rotation of abc, the negative-sequence components have a phase rotation of acb and the zero-sequence components are unidirectional. The transformation from the unbalanced system to the sequence component system leads to a single-phase equivalent of the original unbalanced three-phase circuit and enables simple and straightforward analysis.

In Chapter 2, we introduced Euler's formula

$$e^{j\theta} = \cos\theta + j\sin\theta$$

as a mathematical expression. In the context of this chapter, we will consider $e^{j\theta}$ as an operator that rotates a vector through $\theta°$ (Appendix D). The angle of interest is $120°$ that being the angle between the phase voltages and currents in a three-phase system. We define the operator a, where

$$\begin{aligned}a &= e^{j120} \\ &= \cos 120° + j\sin 120° \\ &= -0.5 + j0.866\end{aligned}$$

---

[1] Although Fortescue conceived the concept of symmetrical components, further development of the technique for practical use is due to C.F. Wagner, R.D. Evans and E.L. Harder of Westinghouse and E. Clarke of GE.

Current Interruption Transients Calculation, First Edition. David F. Peelo.
© 2014 John Wiley & Sons, Ltd. Published 2014 by John Wiley & Sons, Ltd.

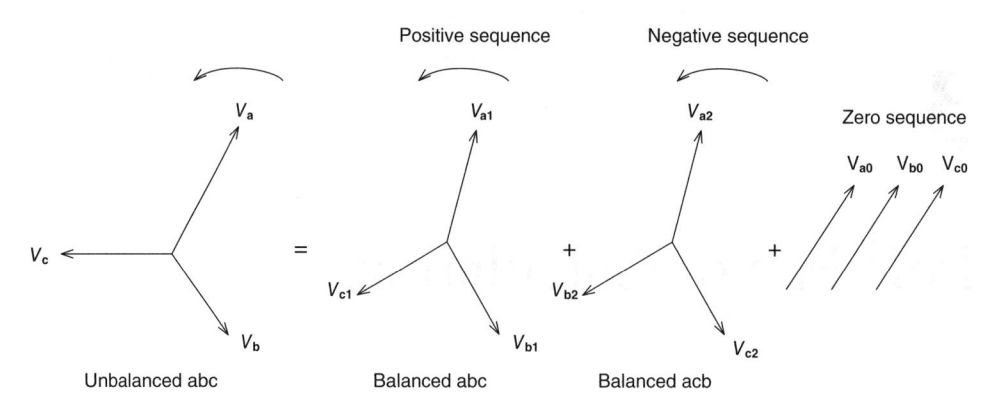

Figure 3.1 Symmetrical component set.

- Unbalanced abc
- Positive sequence — Balanced abc
- Negative sequence — Balanced acb
- Zero sequence

and

$$a^2 = e^{j240}$$
$$= \cos 240° + j \sin 240°$$
$$= -0.5 - j0.866.$$

The operator a can appear in a number of forms, and a summary is provided in Table 3.1. From Figure 3.1, we can write

$$V_a = V_{a0} + V_{a1} + V_{a2},$$
$$V_b = V_{b0} + V_{b1} + V_{b2},$$
$$V_c = V_{c0} + V_{c1} + V_{c2}.$$

Table 3.1 Operator a summary.

Form	Complex number	e-vector
a	$-0.5 + j0.866$	e^{j120}
a^2	$-0.5 - j0.866$	e^{j240}
a^3	$1 + j0$	e^{j0}
$a^4 = a$	$-0.5 + j0.866$	e^{j120}
$a^5 = a^2$	$-0.5 + j0.866$	e^{j240}
$a + a^2 + 1$	0	0
$a + a^2$	$-1 + j0$	e^{j180}
$a - a^2$	$0 + j1.732$	$1.732 e^{j90}$
$a^2 - a$	$0 - j1.732$	$1.732 e^{j270}$
$1 - a$	$1.5 - j0.866$	$1.732 e^{j330}$
$1 - a^2$	$1.5 + j0.866$	$1.732 e^{j30}$
$a - 1$	$-1.5 + j0.866$	$1.732 e^{j150}$
$a^2 - 1$	$-1.5 - j0.866$	$1.732 e^{j210}$
$1 + a$	$0.5 + j0.866$	e^{j60}
$1 + a^2$	$-0.5 + j0.866$	e^{j300}

Pole Factor Calculation

Again from Figure 3.1 and applying the operators a and a^2,

$$V_{b1} = a^2 V_{a1},$$
$$V_{b2} = a V_{a2},$$
$$V_{c1} = a V_{a1},$$
$$V_{c2} = a^2 V_{a2},$$
$$V_{a0} = V_{b0} = V_{c0} = V_0.$$

Designating V_{a1} and V_{a2} now as V_1 and V_2, the positive- and negative-sequence voltages, V_a, V_b and V_c become

$$V_a = V_0 + V_1 + V_2,$$
$$V_b = V_0 + a^2 V_1 + a V_2,$$
$$V_c = V_0 + a V_1 + a^2 V_2.$$

This is written in matrix form as

$$\begin{bmatrix} V_a \\ V_b \\ V_c \end{bmatrix} = \begin{bmatrix} 1 & 1 & 1 \\ 1 & a^2 & a \\ 1 & a & a^2 \end{bmatrix} \begin{bmatrix} V_0 \\ V_1 \\ V_2 \end{bmatrix}. \tag{3.1}$$

Taking the inverse,

$$\begin{bmatrix} V_0 \\ V_1 \\ V_2 \end{bmatrix} = \frac{1}{3} \begin{bmatrix} 1 & 1 & 1 \\ 1 & a & a^2 \\ 1 & a^2 & a \end{bmatrix} \begin{bmatrix} V_a \\ V_b \\ V_c \end{bmatrix}. \tag{3.2}$$

In short form, Eq. (3.2) becomes

$$V_{012} = T \cdot V_{abc}, \tag{3.3}$$

where

$$T = \frac{1}{3} \begin{bmatrix} 1 & 1 & 1 \\ 1 & a & a^2 \\ 1 & a^2 & a \end{bmatrix},$$

and Eq. (3.1) becomes

$$V_{abc} = T^{-1} \cdot V_{012}, \tag{3.4}$$

where

$$T^{-1} = \begin{bmatrix} 1 & 1 & 1 \\ 1 & a^2 & a \\ 1 & a & a^2 \end{bmatrix}.$$

Likewise, for the currents,

$$I_{012} = T \cdot I_{abc} \tag{3.5}$$

and

$$I_{abc} = T^{-1} \cdot I_{012}. \tag{3.6}$$

Similar equations apply for impedances, and the sequence impedances are denoted Z_0, Z_1 and Z_2. The impedances are fixed vectors and do not rotate as the voltage and current vectors will do. Each impedance can be written in the form

$$Z = R + jX,$$

but the R term tends to be less significant than the jX term and is usually ignored in fault current calculations. Therefore, only X_0, X_1 and X_2 will be used in the following. Also, for faults remote from generation, X_1 and X_2 can be taken as equal, and this will be applied in all calculations.

Each sequence has its own network, essentially the Thevenin equivalent circuit for each sequence as seen from the fault point. The sequence networks are shown in Figure 3.2. Only the positive sequence has a voltage source because the system supply is taken as being balanced. V_{af} is the phase voltage to earth at the fault location before the fault. The interconnection of the three networks is dependent on the fault types and can be determined in general by applying Kirchoff's voltage and current laws. Sequence current flow directions should be consistent, that is, as shown in Figure 3.2.

In applying symmetrical components, the basic principle is that of superposition. The fault scenario is viewed in isolation and, if necessary, can be added to other ongoing events. Readers are encouraged to always use a consistent approach such as that shown in Figure 3.3.

By way of example, take the case of a three-phase-to-earth fault:

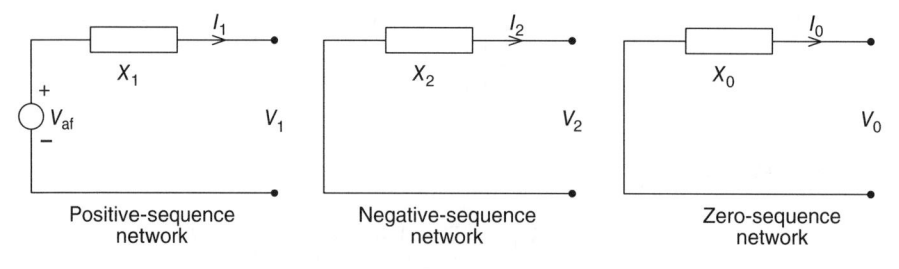

Figure 3.2 Positive-, negative- and zero-sequence networks.

Pole Factor Calculation

```
┌─────────────────────────┐
│ Draw equivalent circuit │
│    for fault case       │
│                    abc  │
└───────────┬─────────────┘
            │
            ▼
┌─────────────────────────┐
│ Write equations for     │
│       fault case        │
│                    abc  │
└───────────┬─────────────┘
           T│
            ▼
┌─────────────────────────┐
│ Write equations for     │
│       fault case        │
│                    012  │
└───────────┬─────────────┘
            │
            ▼
┌─────────────────────────┐
│ Draw equivalent circuit │
│ diagram using sequence  │
│       networks          │
│                    012  │
└───────────┬─────────────┘
            │
            ▼
┌─────────────────────────┐
│ Calculate sequence      │
│ voltages and currents   │
│                    012  │
└───────────┬─────────────┘
         T⁻¹│
            ▼
┌─────────────────────────┐
│ Fault voltages and      │
│      currents           │
│                    abc  │
└─────────────────────────┘
```

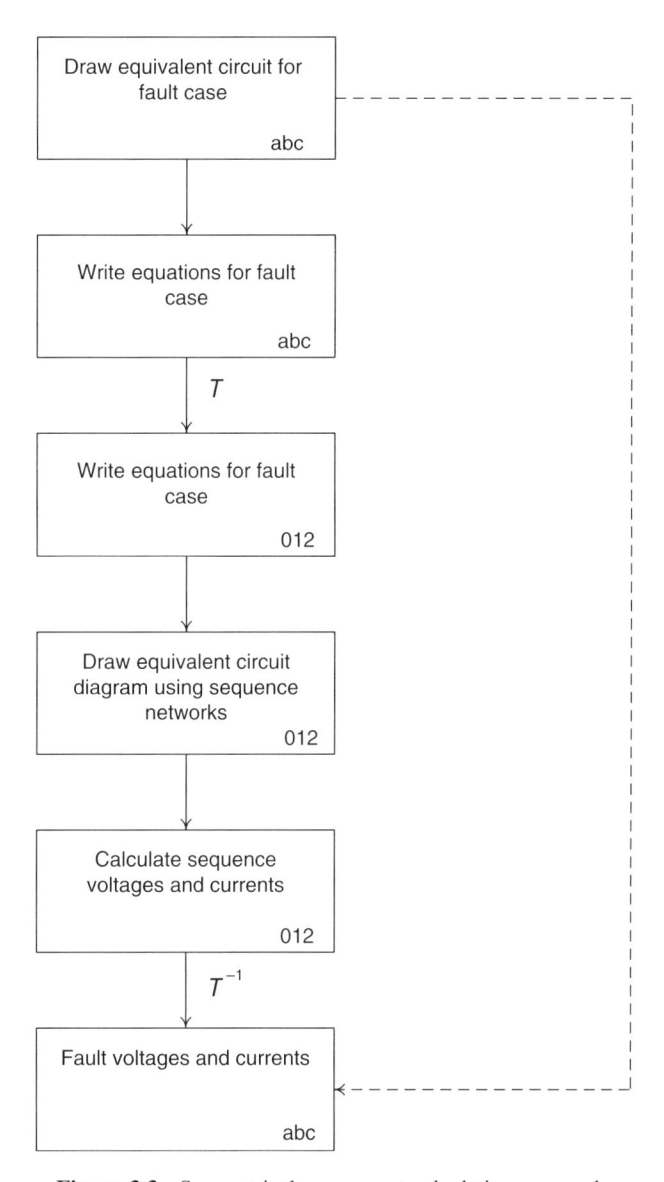

Figure 3.3 Symmetrical component calculation approach.

Step 1: Draw the equivalent circuit for the fault case.
  The equivalent circuit is shown in Figure 3.4, where V_a, V_b and V_c are phase voltages to earth, and I_a, I_b and I_c are the respective fault currents in each phase.

Step 2: Write equations for the fault case in abc notation.
  The fault is balanced, and we can write

$$V_a = V_b = V_c = 0$$

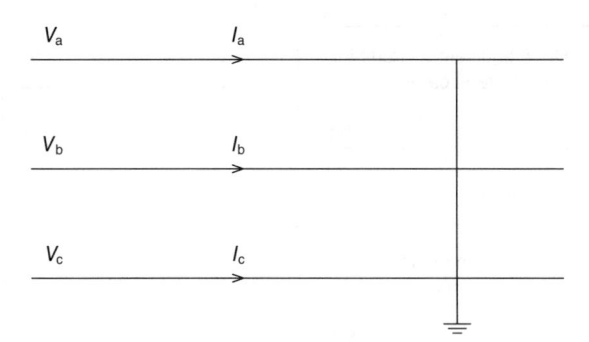

Figure 3.4 Equivalent circuit for a three-phase-to-earth fault.

and
$$I_a + I_b + I_c = 0.$$

Step 3: Write equations for the fault case in 012 notation.

To derive the sequence voltages and currents, we must apply the T transformation (Eq. (3.3)):

$$\begin{bmatrix} V_0 \\ V_1 \\ V_2 \end{bmatrix} = T \begin{bmatrix} V_a \\ V_b \\ V_c \end{bmatrix},$$

giving

$$V_0 = V_1 = V_2 = 0,$$

and from Eq. (3.5),

$$\begin{bmatrix} I_0 \\ I_1 \\ I_2 \end{bmatrix} = T \begin{bmatrix} I_1 \\ I_2 \\ I_3 \end{bmatrix},$$

giving

$$\begin{aligned} I_0 &= 1/3(I_a + I_b + I_c) \\ &= 0, \\ I_1 &= 1/3(I_a + aI_b + a^2 I_c), \end{aligned} \tag{3.7}$$

$$I_2 = 1/3(I_a + a^2 I_b + aI_c). \tag{3.8}$$

Because the fault currents are balanced, we can write

$$I_b = a^2 I_a \quad \text{and} \quad I_c = aI_a$$

Pole Factor Calculation

and substituting in Eqs. (3.7) and (3.8),

$$I_1 = 1/3 I_a (1 + a^3 + a^3)$$
$$= I_a \quad (a^3 = 1 \text{ from Table 3.1}),$$
$$I_2 = 1/3 I_a (1 + a^4 + a^2)$$
$$= 0 \quad (a^4 = a \text{ from Table 3.1}).$$

The sequence equations for the fault case are

$$V_0 = V_1 = V_2 = 0 \tag{3.9}$$

and

$$I_0 = 0,$$
$$I_1 = I_a,$$
$$I_2 = 0.$$

Step 4: Draw an equivalent circuit diagram using sequence networks.

With reference to Figure 3.2, Eq. (3.9) indicates that the sequence network terminals are shorted, and obviously only positive-sequence current can flow as shown in Figure 3.5.

Step 5: Calculate sequence voltages and currents.

Only the positive-sequence current has a value, and from Figure 3.5,

$$I_1 = \frac{V_{af}}{X_1} \quad \text{and} \quad I_0 = I_2 = 0.$$

Step 6: Fault currents and voltages (in abc notation).

Taking the reverse T^{-1} transformation (Eq. (3.6)),

$$\begin{bmatrix} I_a \\ I_b \\ I_c \end{bmatrix} = T^{-1} \begin{bmatrix} 0 \\ I_1 \\ 0 \end{bmatrix},$$

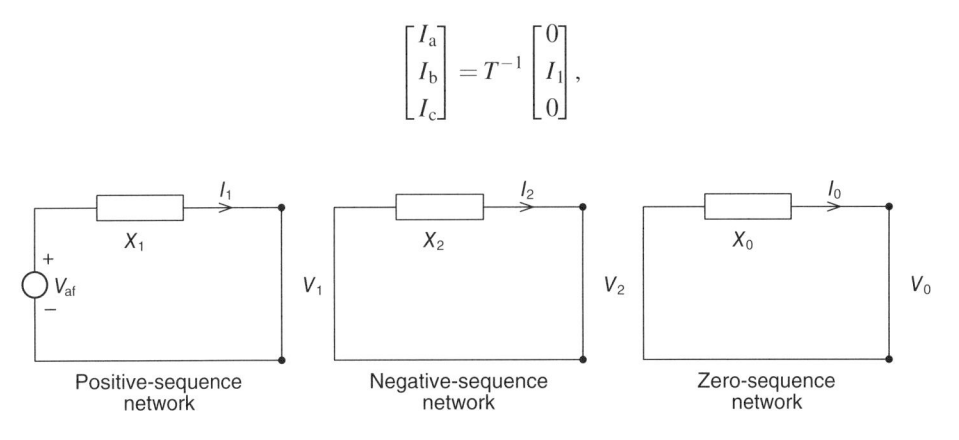

Figure 3.5 Sequence network for a three-phase-to-earth fault.

which gives

$$I_a = I_1,$$
$$I_b = a^2 I_1,$$
$$I_c = a I_1.$$

The phase fault currents are balanced and have a magnitude of V_{af}/X_1, that is, the prefault voltage divided by the positive-sequence impedance.

Although symmetrical components are used more to calculate fault currents, we are interested in the power frequency recovery voltages that occur when circuit breakers interrupt fault currents. Here we need to digress and discuss the current interrupting rating basis for circuit breakers to provide the context for the voltage calculations.

According to IEC 62271-100, the IEC standard for AC circuit breakers rated at 1 kV and above, circuit breakers are applied on either an effectively earthed system or a non-effectively earthed system. An effectively earthed system is defined as one in which $X_0/X_1 \leq 3$ and a non-effectively earthed system as one having isolated, high-impedance or resonant earthed neutral systems.

For circuit breakers applied on effectively earthed systems, the fault current interrupting rating basis is that for a three-phase-to-earth fault. As shown in Figure 3.6, the individual phases will clear in sequence because current zeros occur in the respective phases (this is discussed further in Chapter 4). Referring to Figure 3.6, the sequence of fault current interruption is as follows:

- Initially, all three-phases carry fault current. At some point, the arcing time in one phase will equal or exceed the minimum arcing time, and the current will be interrupted in that phase (actually a-phase in Figure 3.6). The circuit breaker pole in a-phase is referred to as the first-pole-to-clear. The recovery voltage that appears on the source side of the circuit breaker pole in a-phase in turn is usually expressed as a factor k_{pp} in pu, equal to the recovery voltage divided by the prefault voltage and referred to as the first-pole-to-clear factor.
- After the first pole clears, this leaves a two-phase-to-earth fault, usually and hereafter referred to as a double-line-to-earth fault. As long as this condition exists, V_a is the recovery voltage for the first-pole-to-clear, and we can write

$$k_{pp1} = \frac{V_a}{V_{af}}. \tag{3.10}$$

k_{pp1} is the highest of all the pole factors and is the factor listed for transient recovery voltages (TRVs) in the IEC circuit breaker standard and designated only as k_{pp}.
- After the second pole clears, this in turn leaves a single-phase-to-earth fault, usually and hereafter referred to as a single-line-to-earth fault. Again, as long as this condition exists, V_a and V_b give the recovery voltage for the second-pole-to-clear, and we can write

$$k_{pp2} = \frac{V_a}{V_{af}} \tag{3.11}$$

Pole Factor Calculation

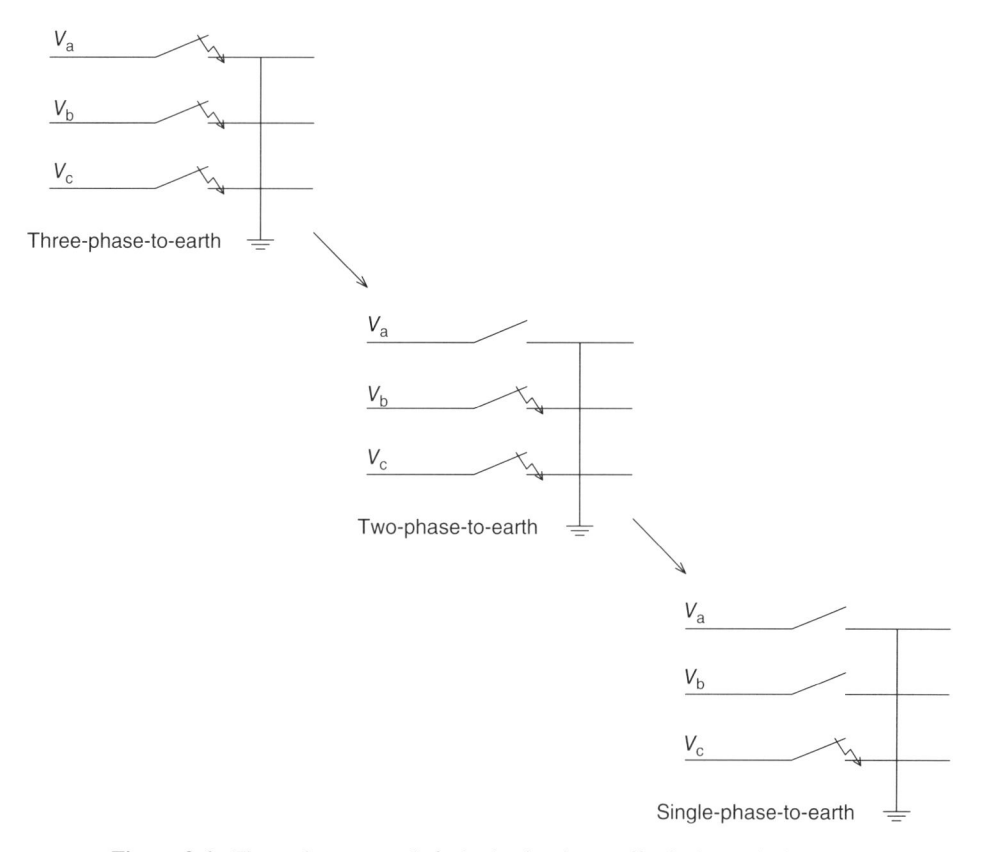

Figure 3.6 Three-phase-to-earth fault clearing in an effectively earthed system.

or

$$k_{pp2} = \frac{V_b}{V_{af}}. \qquad (3.12)$$

V_a and V_b have equal absolute values but are phase shifted from one another.
- When the third pole clears, the third-pole-to-clear recovery voltage is V_c, and we can write

$$k_{pp3} = \frac{V_c}{V_{af}}.$$

However, the system is now balanced, $V_c = V_{af}$ and $k_{pp3} = 1$.

For circuit breakers applied on non-effectively earthed systems, the fault current interrupting rating basis is that for a three-phase unearthed fault. This is shown in Figure 3.7, and the sequence of fault current interruption is similar to that for the above effectively earthed system case:

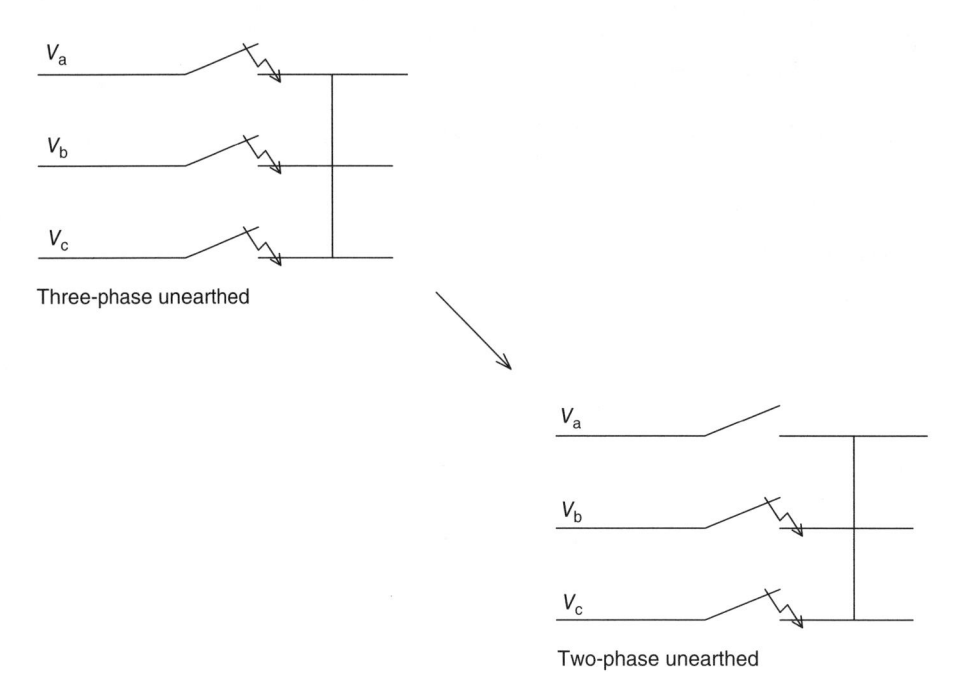

Figure 3.7 Three-phase unearthed fault clearing in a non-effectively earthed system.

- After the first pole clears, this leaves a two-phase fault, usually and hereafter referred to as a line-to-line fault. As long as this condition exists, V_a is the source-side recovery voltage to earth for the first-pole-to-clear, and we can write

$$k_{pps} = \frac{V_a}{V_{af}}. \qquad (3.13)$$

However, because the fault is unearthed, there is a voltage on the neutral point equal to $-0.5 V_{af}$, and the first-pole-to-clear recovery voltage will be the difference between source-side and neutral voltages.
- After first pole clearing, the fault currents in b-phase and c-phase will each phase shift by 30° to become equal in magnitude and opposite in polarity. The second and third poles will clear simultaneously in series against a line-to-line voltage.

When using symmetrical components to calculate fault currents, the equivalent circuit for the fault case is obviously that when the currents are still flowing, that is, before current interruption. To calculate recovery voltage, the equivalent circuit is one step down, that is, the current has been interrupted in the phase of interest while the fault currents in the other phases continue to persist or has been interrupted as in the case of a single-line-to-earth fault. Therefore, the cases that we need to consider are the following:

1. Effectively earthed systems.
    - Double-line-to-earth fault.
    - Single-line-to-earth fault.

2. Non-effectively earthed systems.
   - Line-to-line fault.

3.2 Pole Factors: Effectively Earthed Systems

The first case to consider is the double-line-to-earth fault as shown in Figure 3.8. The intent is to calculate the value of V_a.

The fault equations are

$$I_a = 0,$$
$$V_b = V_c = 0.$$

Applying the transformation T,

$$\begin{bmatrix} V_0 \\ V_1 \\ V_2 \end{bmatrix} = T \begin{bmatrix} V_a \\ 0 \\ 0 \end{bmatrix},$$

giving

$$V_0 = V_1 = V_2 = \frac{1}{3} V_a \qquad (3.14)$$

and

$$\begin{bmatrix} I_0 \\ I_1 \\ I_2 \end{bmatrix} = T \begin{bmatrix} 0 \\ I_b \\ I_c \end{bmatrix},$$

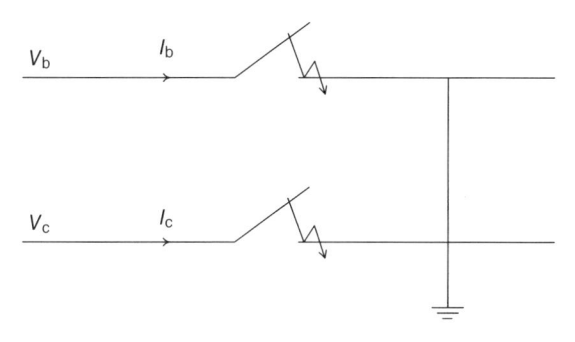

Figure 3.8 Equivalent circuit for a double-line-to-earth fault.

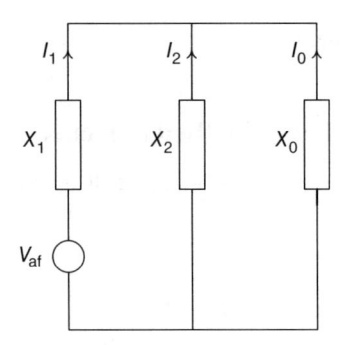

Figure 3.9 Sequence network for a double-line-to-earth fault.

$$I_0 = \frac{1}{3}(I_b + I_c),$$
$$I_1 = \frac{1}{3}(a^2 I_b + a I_c),$$
$$I_2 = \frac{1}{3}(a I_b + a^2 I_c)$$

and

$$I_0 + I_1 + I_2 = 0. \tag{3.15}$$

Equations (3.14) and (3.15) indicate that the positive-, negative- and zero-sequence networks are connected in parallel as shown in Figure 3.9. From Eq. (3.14), we can write

$$V_1 = \frac{1}{3} V_a,$$

and from Figure 3.9,

$$V_1 = V_{af} - X_1 I_1$$

and

$$I_1 = \frac{V_{af}}{X_1 + (X_2 X_0 / (X_2 + X_0))},$$

$$V_a = 3 V_{af} \left(1 - \frac{X_1}{X_1 + (X_2 X_0 / (X_2 + X_0))}\right)$$

$$= 3 V_{af} \left(\frac{X_2 X_0}{X_1 X_2 + X_1 X_0 + X_2 X_0}\right).$$

From Eq. (3.10),

$$k_{pp1} = \frac{V_a}{V_{af}}$$

$$= \frac{3 X_2 X_0}{X_1 X_2 + X_1 X_0 + X_2 X_0}.$$

Pole Factor Calculation

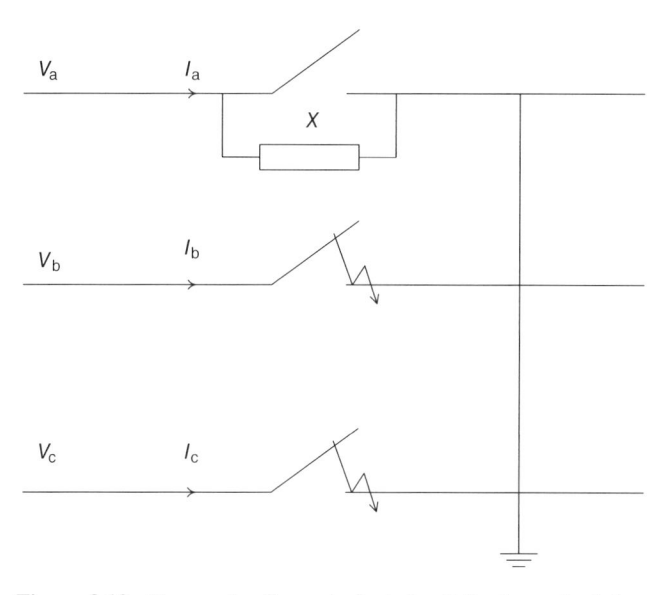

Figure 3.10 Hammarlund's equivalent circuit for k_{pp1} calculation.

For transmission systems, X_1 and X_2 are taken as being equal, and

$$k_{pp1} = \frac{3X_0}{X_1 + 2X_0}. \tag{3.16}$$

Equation (3.16) is often written as

$$k_{pp1} = \frac{3(X_0/X_1)}{1 + 2(X_0/X_1)} \tag{3.17}$$

and for $X_0/X_1 = 3$ (for effectively earthed systems), $k_{pp1} = 1.286$ pu. For standardization purposes, this value is rounded up to 1.3 pu.

Hammarlund proposed that the k_{pp1} factor could also be calculated by adding an impedance X across the first-pole-to-clear as shown in Figure 3.10.[2] The value of V_a is again calculated, and then X is allowed to go to infinity.

The fault equations are

$$V_a = XI_a,$$
$$V_b = V_c = 0.$$

---

[2] Hammarlund's report (see Bibliography at the end of this chapter and Appendix H) is without doubt the most quoted reference ever with respect to pole factors. It is, however, strange to note that although he made the proposal, he did not perform the calculations to verify if the proposal is correct or not.

Applying the transformation T,

$$\begin{bmatrix} V_0 \\ V_1 \\ V_2 \end{bmatrix} = T \begin{bmatrix} XI_a \\ 0 \\ 0 \end{bmatrix},$$

giving

$$V_0 = V_1 = V_2 = \frac{1}{3} XI_a$$

or

$$3V_0 = 3V_1 = 3V_2 = XI_a, \qquad (3.18)$$

$$\begin{bmatrix} I_0 \\ I_1 \\ I_2 \end{bmatrix} = T \begin{bmatrix} I_a \\ I_b \\ I_c \end{bmatrix},$$

$$\begin{aligned} I_0 &= \frac{1}{3}(I_a + I_b + I_c), \\ I_1 &= \frac{1}{3}(I_a + aI_b + a^2 I_c), \\ I_2 &= \frac{1}{3}(I_a + a^2 I_b + a I_c), \\ I_0 + I_1 + I_2 &= I_a. \end{aligned} \qquad (3.19)$$

Equations (3.18) and (3.19) indicate a parallel connection of the positive-, negative- and zero-sequence networks and that of the load circuit represented by the impedance X. This connection is shown in Figure 3.11, and we can note that V_{af} and the sequence impedances have to be multiplied by the factor 3 of Eq. (3.18) to meet the requirements of Eq. (3.19).

As shown in Figure 3.11, the calculation is simplified by considering the Thevenin equivalent circuit. To calculate the Thevenin equivalent voltage V_{th}, we remove the load

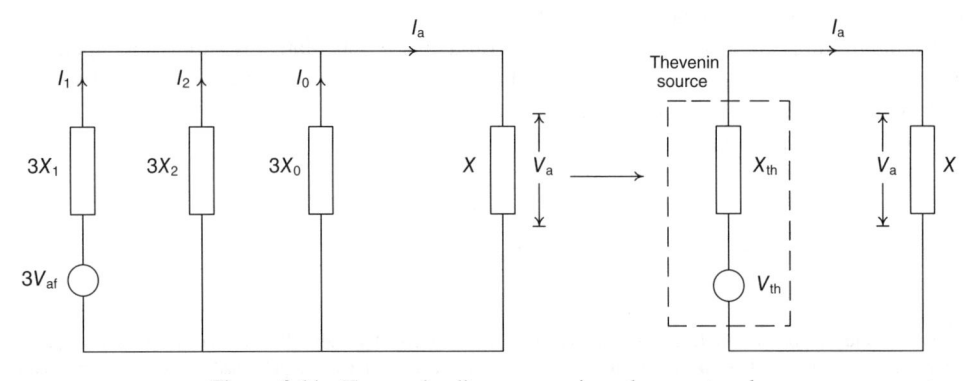

Figure 3.11 Hammarlund's sequence impedance network.

Pole Factor Calculation

and derive the voltage seen by the load. We can see immediately that the remaining circuit is similar to that of Figure 3.9, and we can write

$$I_1 = \frac{3V_{af}}{3X_1 + (3X_2X_0/(X_2+X_0))}$$
$$= \frac{V_{af}(X_2+X_0)}{X_1X_2 + X_1X_0 + X_2X_0}.$$

Again, taking $X_1 = X_2$,

$$V_{th} = \frac{V_{af}(X_1+X_0)}{X_1^2 + 2X_1X_0} \cdot \frac{3X_1X_0}{X_1+X_0}$$
$$= \frac{3V_{af}X_0}{X_1 + 2X_0}. \quad (3.20)$$

This equation is obviously familiar, and from Figure 3.11,

$$V_a = XI_a$$
$$= V_{th} \cdot \frac{X}{X_{th} + X}$$
$$= V_{th} \cdot \frac{1}{X_{th}/X + 1}.$$

At $X \to \infty$, $V_{th} = V_a$ and Eq. (3.20) becomes

$$\frac{V_a}{V_{af}} = \frac{3X_0}{X_1 + 2X_0},$$

identical to Eq. (3.16) and proving Hammarlund's proposal.

The recovery voltage on the second-pole-to-clear is calculated using the fault circuit shown in Figure 3.12. This voltage exists only for the time between first pole and third pole fault current interruption. As noted earlier, the recovery voltage appears on both a- and b-phases.[3]

The fault equations are

$$V_c = 0,$$
$$I_a = I_b = 0.$$

---

[3] Any phase can be taken as the reference phase provided that the phase sequence is maintained.

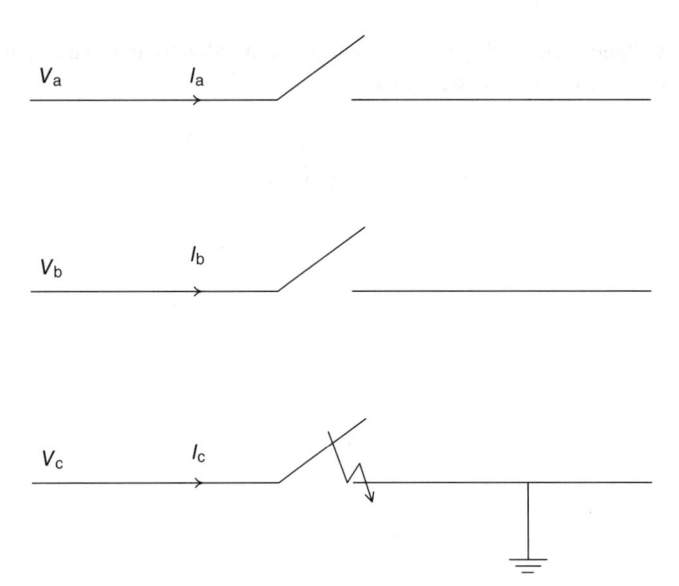

Figure 3.12 Equivalent circuit for a single-line-to-earth fault.

Applying the transformation T,

$$\begin{bmatrix} V_0 \\ V_1 \\ V_2 \end{bmatrix} = T \begin{bmatrix} 0 \\ V_a \\ V_b \end{bmatrix},$$

$$V_0 = \frac{1}{3}(V_a + V_b),$$

$$V_1 = \frac{1}{3}(aV_a + a^2 V_b),$$

$$V_2 = \frac{1}{3}(a^2 V_a + aV_b),$$

$$V_0 + V_1 + V_2 = 0,$$

$$\begin{bmatrix} I_0 \\ I_1 \\ I_2 \end{bmatrix} = T \begin{bmatrix} I_c \\ 0 \\ 0 \end{bmatrix},$$

$$I_0 = I_1 = I_2 = \frac{1}{3}I_c. \tag{3.22}$$

Equations (3.21) and (3.22) indicate that the positive-, negative- and zero-sequence networks are connected in series as shown in Figure 3.13.

Pole Factor Calculation

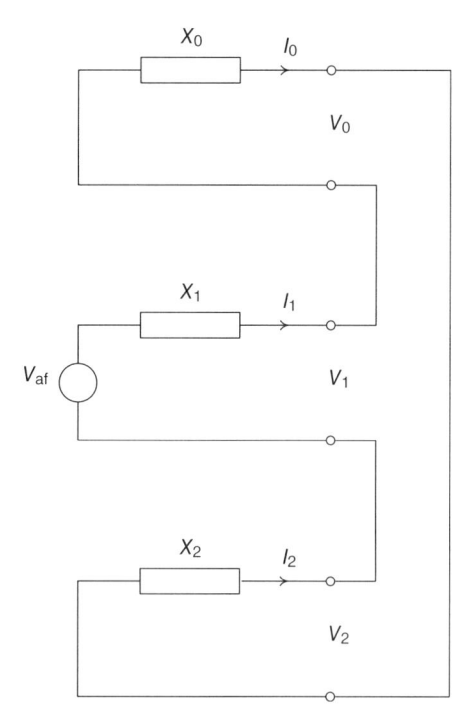

Figure 3.13 Sequence network for a single-line-to-earth fault.

From Figure 3.12, we can write

$$I_0 = I_1 = I_2$$
$$= \frac{V_{af}}{X_0 + X_1 + X_2},$$
$$V_0 = -X_0 I_0$$
$$= -\frac{X_0 V_{af}}{X_0 + X_1 + X_2},$$
$$V_1 = -X_1 I_1 + V_{af}$$
$$= V_{af}\left(1 - \frac{X_1}{X_0 + X_1 + X_2}\right)$$
$$= V_{af}\left(\frac{X_0 + X_2}{X_0 + X_1 + X_2}\right),$$
$$V_2 = -X_2 I_2$$
$$= -\frac{X_2 V_{af}}{X_0 + X_1 + X_2}.$$

Taking the inverse transformation T^{-1},

$$\begin{bmatrix} V_c \\ V_a \\ V_b \end{bmatrix} = T^{-1} \begin{bmatrix} V_0 \\ V_1 \\ V_2 \end{bmatrix},$$

$V_c = 0,$

$V_a = V_0 + a^2 V_1 + a V_2$

$$= V_{af}\left(-\frac{X_0}{X_0 + X_1} X_2 + a^2 - \frac{a^2 X_1}{X_0 + X_1 + X_2} - \frac{a X_2}{X_0 + X_1 + X_2} \right) \quad (3.23)$$

$$= V_{af}\left(a^2 - \frac{X_0 + a^2 X_1 + a X_2}{X_0 + X_1 + X_2} \right),$$

$V_b = V_0 + a V_1 + a^2 V_2$

$$= V_{af}\left(a - \frac{X_0 + a X_1 + a^2 X_2}{X_0 + X_1 + X_2} \right). \quad (3.24)$$

Taking $X_1 = X_2$, from Eqs. (3.23) and (3.24),

$$\frac{V_a}{V_{af}} = a^2 - \frac{X_0 + (a + a^2) X_1}{X_0 + 2 X_1}$$

$$= \frac{1}{X_0 + 2 X_1}\left[(a^2 - 1) X_0 + (a^2 - a) X_1 \right]$$

and with reference to Table 3.1

$$\frac{V_a}{V_{af}} = \frac{1}{X_0 + 2 X_1}\left[-1.5 X_0 - j\sqrt{3}(0.5 X_0 + X_1) \right]. \quad (3.25)$$

Similarly, readers can show that

$$\frac{V_b}{V_{af}} = \frac{1}{X_0 + 2 X_1}\left[-1.5 X_0 + j\sqrt{3}(0.5 X_0 + X_1) \right]. \quad (3.26)$$

Clearly, V_a/V_{af} and V_b/V_{af} are equal in magnitude but phase shifted from one another. To now calculate k_{pp2}, we take the absolute values of Eqs. (3.25) and (3.26):

$$k_{pp2} = \left| \frac{V_a}{V_{af}} \right|$$

$$= \left| \frac{V_b}{V_{af}} \right|$$

$$= \frac{1}{X_0 + 2 X_1} \sqrt{3(X_0^2 + X_0 X_1 + X_1^2)}$$

most often written in the form

$$k_{pp2} = \frac{\sqrt{3}}{2 + X_0/X_1}\sqrt{1 + \frac{X_0}{X_1} + \left(\frac{X_0}{X_1}\right)^2}. \quad (3.27)$$

Taking $X_0/X_1 = 3$ in Eq. (3.27) for an effectively earthed system, $k_{pp2} = 1.249$ pu. For standardization purposes, k_{pp2} is taken as 1.27 pu.[4]

After the third pole clears, the system is again balanced and $k_{pp3} = 1$. Readers are encouraged to prove this statement using Hammarlund's approach, that is, apply an impedance X across the third pole and then allow it to go to infinity.

3.3 Pole Factors: Non-Effectively Earthed Systems

The first case to consider is the line-to-line fault shown in Figure 3.14. The intent is to first calculate the value of V_a and then the value of the voltage on the neutral point. The difference between these two voltages is the first-pole-to-clear recovery voltage.

The fault equations are

$$I_a = 0,$$
$$V_b = V_c,$$
$$I_c = -I_b.$$

Applying the transformation T,

$$\begin{bmatrix} V_0 \\ V_1 \\ V_2 \end{bmatrix} = T \begin{bmatrix} V_a \\ V_b \\ V_c \end{bmatrix},$$

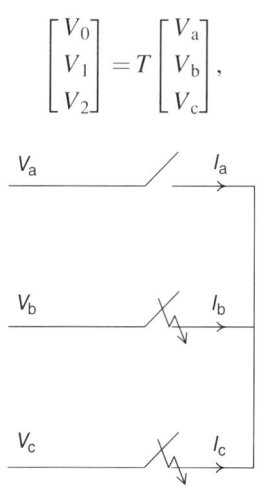

Figure 3.14 Equivalent circuit for a line-to-line fault.

---

[4] The reason for the difference between the calculated and the standard value is not documented. A possible explanation is as follows. If we calculate the value of X_0/X_1 using Eq. (3.17) and use the rounded-up value of 1.3 for k_{pp1}, we get $X_0/X_1 = 3.25$ pu. Substitution of this value in Eq. (3.28) gives $k_{pp2} = 1.27$ pu.

$$V_0 = \frac{1}{3}(V_a + V_b + V_c)$$
$$= 0 \quad (\text{why?}),$$
$$V_1 = \frac{1}{3}(V_a + aV_b + a^2V_c), \tag{3.28}$$
$$V_2 = \frac{1}{3}(V_a + a^2V_b + aV_c),$$
$$V_2 = V_1 \quad \text{because } V_b = V_c,$$

$$\begin{bmatrix} I_0 \\ I_1 \\ I_2 \end{bmatrix} = T \begin{bmatrix} 0 \\ I_b \\ -I_b \end{bmatrix},$$

$$\begin{aligned} I_0 &= 0, \\ I_1 &= \frac{1}{3}(aI_b - a^2 I_b), \\ I_2 &= \frac{1}{3}(a^2 I_b - aI_b), \\ I_2 &= -I_1. \end{aligned} \tag{3.29}$$

Equations (3.28) and (3.29) indicate that the positive- and negative-sequence networks are connected in parallel as shown in Figure 3.15. Note that the zero-sequence network is not connected but short-circuited because $V_0 = 0$ and in effect does not exist.

From Figure 3.15, we can write

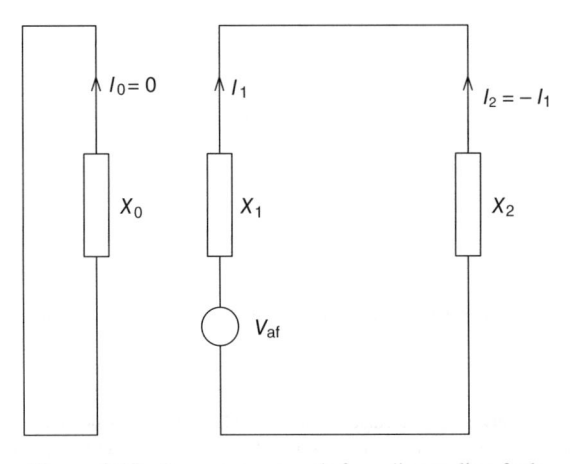

$$\begin{aligned} V_1 &= -X_1 I_1 + V_{af} \\ &= -X_1 \frac{V_{af}}{X_1 + X_2} + V_{af}. \end{aligned}$$

Figure 3.15 Sequence network for a line-to-line fault.

Pole Factor Calculation

Taking $X_1 = X_2$,

$$V_1 = 0.5 V_{af}.$$

Applying the reverse transformation T^{-1},

$$\begin{bmatrix} V_a \\ V_b \\ V_c \end{bmatrix} = T^{-1} \begin{bmatrix} 0 \\ V_1 \\ V_1 \end{bmatrix}, \qquad (3.30)$$

$$\begin{aligned} V_a &= 2V_1 \\ &= V_{af}, \\ V_a/V_{af} &= 1. \end{aligned}$$

Therefore, the recovery voltage on a-phase on the source side of the circuit breaker is 1 pu, which is the expected result because the source remains balanced. However, there is also a first-pole recovery voltage on the neutral point that exists for the time between a-phase current interruption and the simultaneous current interruption in b-phase and c-phase. To calculate the neutral point voltage, we need to determine the value of either V_b or V_c.

From Eq. (3.30), we can write

$$\begin{aligned} V_b &= (a^2 + a) V_1 \\ &= -V_1 \quad \text{(refer to Table 3.1)} \\ &= -0.5 V_{af}. \end{aligned}$$

The recovery voltage V_{ab} across the circuit breaker is given by

$$\begin{aligned} V_{ab} &= V_a - (-0.5 V_{af}) \\ &= 1.5 V_{af}. \end{aligned}$$

The first-pole-to-clear factor k_{pp1} for this case is then

$$\begin{aligned} k_{pp1} &= \frac{V_{cb}}{V_{af}} \\ &= 1.5 \text{ pu}. \end{aligned}$$

After the current is interrupted in the first circuit breaker pole, the currents in b-phase and c-phase will phase shift by 30° either way to become equal in magnitude and opposite in polarity. To calculate the values of V_b and V_c after the current is interrupted, we will use Hammarlund's approach as shown in Figure 3.16.

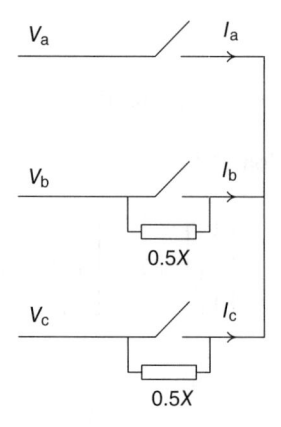

Figure 3.16 Equivalent circuit for line-to-line fault second and third pole factor calculation.

The fault equations are
$$I_a = 0,$$
$$I_c = -I_b,$$
$$V_b - V_c = XI_b.$$

Applying the transformation T,
$$\begin{bmatrix} V_0 \\ V_1 \\ V_2 \end{bmatrix} = T \begin{bmatrix} V_a \\ V_b \\ V_c \end{bmatrix},$$

$$V_0 = \frac{1}{3}(V_a + V_b + V_c)$$
$$= 0 \quad \text{(since the source is balanced)}$$
$$= \frac{1}{3}(V_a + aV_b + a^2 V_c),$$
$$V_1 = \frac{1}{3}\left[V_a + aV_b + a^2(V_b - XI_b)\right]$$
$$= \frac{1}{3}(V_a - V_b - a^2 XI_b).$$

Likewise,
$$V_2 = \frac{1}{3}(V_a - V_b - aXI_b)$$

and
$$V_1 - V_2 = \frac{1}{3}(a - a^2)XI_b,$$

giving
$$V_1 - V_2 = \frac{1}{\sqrt{3}}jXI_b, \tag{3.31}$$

$$\begin{bmatrix} I_0 \\ I_1 \\ I_2 \end{bmatrix} = T \begin{bmatrix} 0 \\ I_b \\ -I_b \end{bmatrix},$$

Pole Factor Calculation

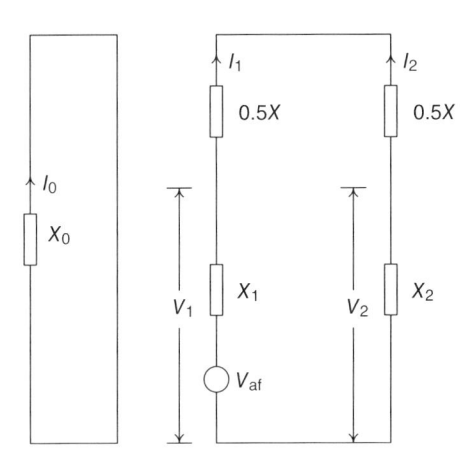

Figure 3.17 Sequence network for line-to-line fault second and third pole factor calculation.

$$I_0 = 0,$$
$$I_1 = \frac{1}{3}(a - a^2)I_b$$
$$= \frac{1}{\sqrt{3}} jI_b,$$
$$I_2 = \frac{1}{3}(a^2 - a)I_b$$
$$= -I_1.$$

From Eq. (3.31), we can write

$$V_1 - V_2 = \frac{1}{\sqrt{3}} jX \frac{\sqrt{3}I_1}{j}$$
$$= XI_1.$$

The calculation indicates a parallel connection of the positive- and negative-sequence networks through an impedance X as shown in Figure 3.17.

Taking $X_1 = X_2$,

$$I_1 = \frac{V_{af}}{2X_1 + X},$$

$$I_b = \frac{\sqrt{3}}{j} \cdot I_1$$
$$= -\frac{j\sqrt{3}V_{af}}{2X_1 + X}.$$

From the fault equations,

$$V_b - V_c = XI_b$$
$$= -\frac{j\sqrt{3}V_{af}}{2(X_1/X) + 1}$$
$$= -j\sqrt{3}V_{af} \quad \text{as } X \to \infty.$$

The recovery voltage for each pole in pu will be

$$\left|\frac{0.5(V_b - V_c)}{V_{af}}\right| = 0.866 \text{ pu},$$

being one-half of the line-to-line voltage.

3.4 Alternative Pole Factor Calculation Method

The first-pole-to-clear calculation for shunt reactor switching, as described in Appendix F, can also be used to calculate the first pole factors for effectively and non-effectively earthed systems. The basic three-phase circuit is shown in Figure F.1 and reproduced here as Figure 3.18.

From Eqs. (F.4) and (F.5), we can write

$$k_{pp1} = \frac{3[X_0 X_1 + 2Z_e X_1 + Z(Z + X_1 + X_0 + 2Z_e)]}{(X_1 + Z)(3Z + X_1 + 2X_0 + 6Z_e)} \tag{3.32}$$

and

$$K = -\frac{-3(X_1 + Z)Z_e}{(X_1 + Z)(3Z + X_1 + 2X_0 + 6Z_e)}. \tag{3.33}$$

For an effectively earthed system and with reference to Figure 3.18,

$$X_1 = X_2, \quad X_0/X_1 = 3, \quad Z = 0, \quad Z_e = 0,$$

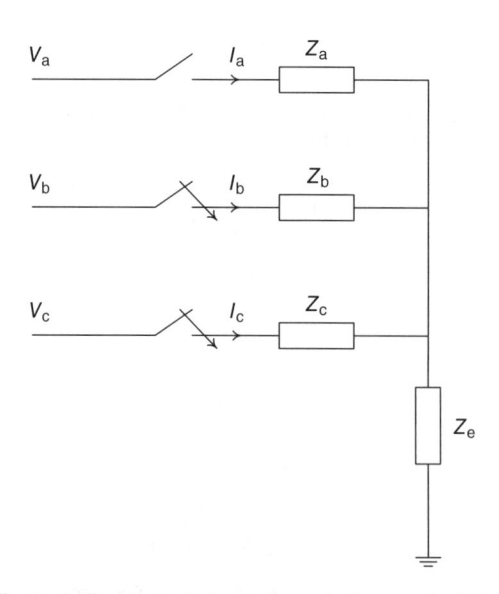

Figure 3.18 General circuit for pole factor calculation.

… Pole Factor Calculation

and applying Eq. (3.32) gives

$$k_{pp1} = \frac{3[X_1X_0 + 0 + 0]}{X_1^2 + 2X_1X_0}$$

$$= \frac{3(X_0/X_1)}{1 + 2(X_0/X_1)} \qquad (3.34)$$

$$= 1.2857 \text{ pu.}$$

For the non-effectively earthed case and again referring to Figure 3.18,

$$X_1 = X_2, \quad X_0 = 0, \quad Z = 0, \quad Z_e = \infty,$$

$$k_{pp1} = \frac{3(0 + 2Z_eX_1 + 0)}{(X_1 + 0)(0 + X_1 + 0 + 2Z_e)}$$

$$= \frac{6Z_eX_1}{X_1^2 + 6X_1Z_e}$$

$$= \frac{6X_1}{X_1^2/Z_e + 6X_1}$$

$$= 1 \text{ pu} \quad \text{as } Z_e \to \infty,$$

$$K = -\frac{3(X_1 + 0)Z_e}{X_1^2 + 6X_1Z_e}$$

$$= -\frac{3X_1}{X_1^2/Z_e + 6X_1}$$

$$= -0.5 \text{ pu} \quad \text{as } Z_e \to \infty.$$

For this case in the literature, it is not uncommon for X_0 to be taken as infinite and Eq. (3.34) is used to derive a k_{pp1} value equal to 1.5 pu. The fact that the numerical solution is correct is coincidental, and this approach is incorrect. From Figure 3.15, the zero-sequence network does not exist, that is, V_0, I_0 and Z_0 all equal zero, and the misconception may arise from the notion that if $I_0 = 0$, then Z_0 must be equal to infinity. A further note is that, taking $X_0 = \infty$ instead of zero, readers can readily determine using Eqs. (3.32) and (3.33) that this gives k_{pp1} and K values of 1.125 and 0.375 pu, respectively, which are, of course, incorrect.

3.5 Three-Phase Test Circuit Arrangement

For circuit breakers applied on effectively earthed systems, short-circuit testing is usually done on a single-phase basis as discussed further in Chapter 4. However, it is possible to derive a laboratory circuit that will give the correct first-pole-to-clear recovery voltage for this case. The circuit, actually the three-phase sequence network, is shown in Figure 3.19, where X_1 is calculated from the short-circuit current level and X_N is the required neutral impedance to give the correct first-pole-to-clear factor. The circuit follows from the calculation in Section 3.1.

From the single-phase circuit to the right in Figure 3.19, which will give the correct recovery voltage and current, using the current injection approach, we write

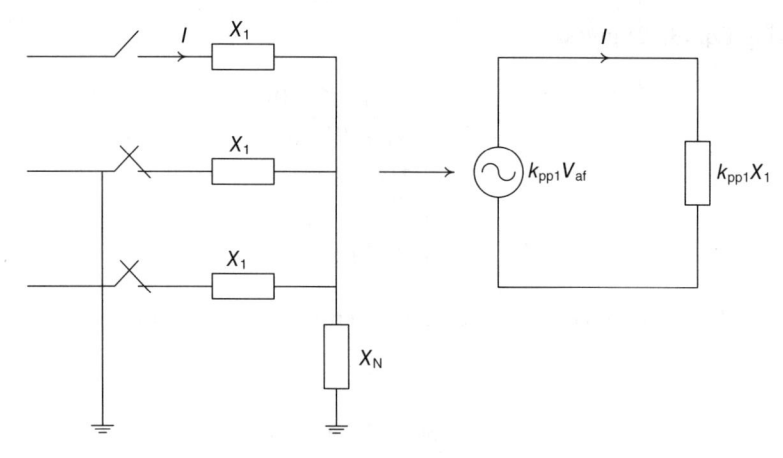

Figure 3.19 Three-phase sequence network for first pole clearing in an effectively earthed system.

$$k_{pp1}X_1 = X_1 + \frac{X_1 X_N}{X_1 + 2X_N}$$
$$= X_1\left(1 + \frac{X_N}{X_1 + 2X_N}\right),$$

giving

$$k_{pp1} = \frac{X_1 + 3X_N}{X_1 + 2X_N}$$
$$= \frac{3X_0}{X_1 + 2X_0} \quad \text{from Eq.(3.34)}.$$

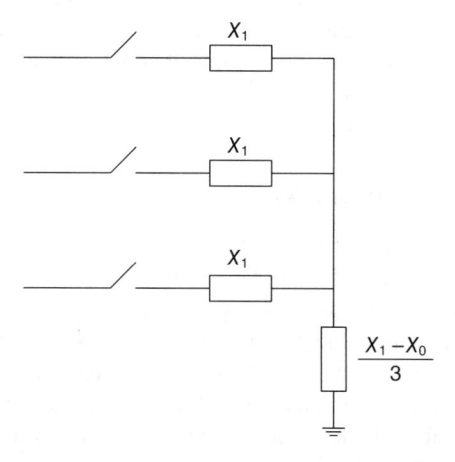

Figure 3.20 Three-phase test circuit for first pole clearing in an effectively earthed system.

Table 3.2 Summary of pole factor calculations.

System earthing	Fault type	Pole factor (pu)		
		First pole	Second pole	Third pole
Effective	Three-phase-to-earth	1.3	1.27	1
Non-effective	Three-phase unearthed	1.5	0.866	0.866

Solving for X_N gives

$$X_N = \frac{1}{3}(X_0 - X_1),$$

and the test circuit is as shown in Figure 3.20. The circuit will give the correct pole factor for first and third pole clearing but not for second pole clearing.

3.6 Summary

The results of the pole factor calculations are summarized in Table 3.2.

The relevance of pole factors can be understood by considering Figure 2.8. The generic curves are characterized by having a starting point (zero), an aiming point or axis of oscillation (1 pu) and a peak value. The pole factor is used to determine the value of the axis of oscillation:

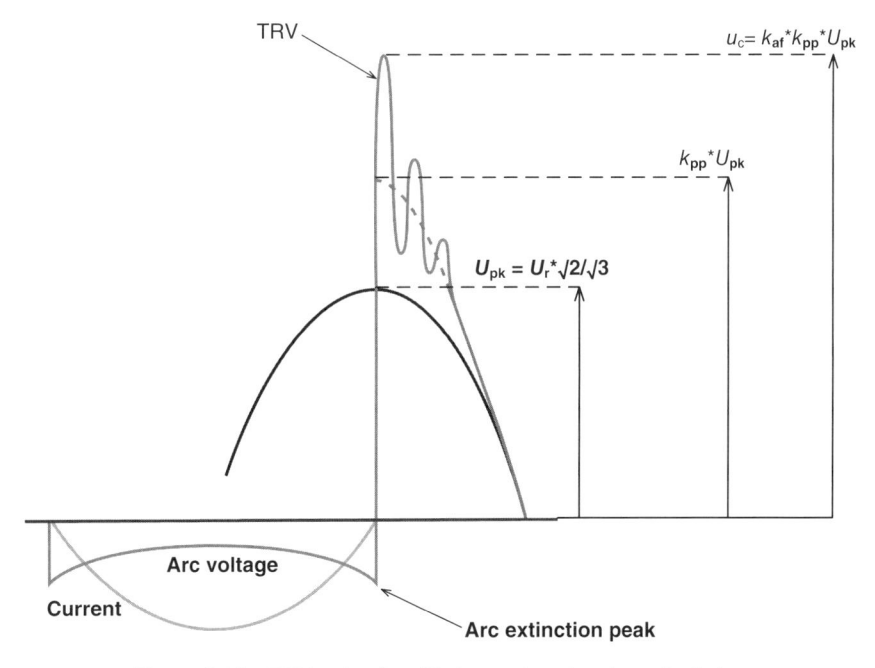

Figure 3.21 TRV axis of oscillation and peak value calculation.

for a three-phase-to-earth fault in an effectively earthed system, 1 pu in Figure 2.8 represents the peak value of the prefault system voltage multiplied by 1.3, which in turn is multiplied by the amplitude factor to determine the TRV peak value. This concept is shown in Figure 3.21 and explored further in the next chapter.

Bibliography

1. Fortescue, C.L. (1918) Method of symmetrical coordinates applied to the solution of polyphase networks. *AIEE Trans.*, **37**, 1027–1140.
2. Wagner, C.F. and Evans, R.D. (1933) *Symmetrical Components*, McGraw-Hill.
3. Clarke, E. (1943) *Circuit Analysis of A-C Power Systems, Volume 1: Symmetrical and Related Components*, John Wiley & Sons, Inc.
4. Blackburn, J.L. (1993) *Symmetrical Components for Power Systems Engineering*, Marcel Dekker.
5. Tleis, N. (2008) *Power System Modelling and Fault Analysis: Theory and Practice*, Elsevier.

4

Terminal Faults

4.1 General Considerations

To this point we have explored the concepts of oscillatory circuits and pole factors. We will now relate these concepts to real transient recovery voltages (TRVs). The TRV requirements for a 245 kV circuit breaker, as specified in IEC 62271-100, are given in Table 4.1.

The table is read as follows by column:

- Column 1: Circuit breaker rated voltage U_r, which is always a maximum voltage.
- Column 2: Short-circuit test duties to which the circuit breaker is type tested; T100 represents a test at the short-circuit rating of the circuit breaker, T60 is a test at 60% of the above rating and so on for T30 and T10. As discussed in detail in Section 4.2, the varying percentage values represent different power system conditions with respect to fault current.
- Column 3: First-pole-to-clear factor k_{pp} as calculated in Chapter 3 for circuit breakers applied on effectively earthed systems. Test duty T10, which represents the case of fault current fed through a transformer, is an exception from the 1.3 pu value. The reason for this is that the pole factor is now dependent on the sequence impedances of a typical power transformer.
- Column 4: Amplitude factors k_{af} as calculated in Chapter 2. The actual values are based on experience, studies and measurements over a long period of time. T10 is again an exception; the amplitude factor is dependent on the low-loss damping RLC characteristics of power transformers. The reason for the 0.9 multiplier is that 90% of the TRV is taken as being associated with the transformer—and experienced by the circuit breaker—and 10% with the source impedance as shown in Figure 4.1.
- Columns 5–8: The columns describe the voltage and time characteristics of the TRVs for the different test duties. The characteristics are discussed in detail in Section 4.2 being overdamped oscillations for T100 and T60 and underdamped oscillations for T30 and T10.
- Columns 9–11: The time delay t_d is a measure of the rate at which the TRV "takes off" immediately after current interruption. As also discussed in detail in Section 4.2, u' and t' describe a delay line, and the allowable excursion of the TRV is limited by this line.
- Column 12: The rate of rise of the TRV, usually designated as the RRRV, is defined by a voltage at a specific time dependent on the test duty and type of oscillation. The RRRV is discussed further in Section 4.2.

Current Interruption Transients Calculation, First Edition. David F. Peelo.
© 2014 John Wiley & Sons, Ltd. Published 2014 by John Wiley & Sons, Ltd.

Table 4.1 IEC TRV requirements for 245 kV circuit breakers.

Rated voltage, U_r (kV)	Test duty	First-pole-to-clear factor, k_{pp} (pu)	Amplitude factor, k_{af} (pu)	First reference voltage, u_1 (kV)	Time, t_1 (µs)	TRV peak value, u_c (kV)	Time, t_2 or t_3 (µs)	Time delay, t_d (µs)	Voltage, u' (kV)	Time, t' (µs)	Rate of rise, u_1/t_1, u_c/t_3 (kV/µs)
245	T100	1.3	1.40	195	98	364	392	2 (27)	98	51 (76)	2
	T60	1.3	1.50	195	65	390	390	2–20	98	35–52	3
	T30	1.3	1.54	—	—	400	80	12	133	39	5
	T10	1.5	0.9 × 1.7	—	—	459	66	10	153	32	7

Terminal Faults

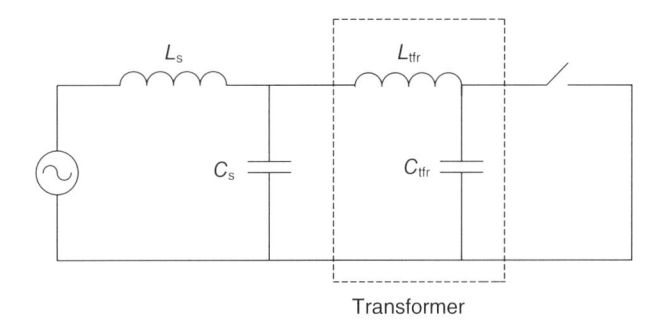

Figure 4.1 Distribution of T10 TRV between the source impedance and transformer.

The IEC TRV requirements for 72.5 kV circuit breakers are given in Table 4.2. In contrast to the 245 kV circuit breaker case, the first-pole-to-clear factors are all 1.5 pu because the power system is taken as being non-effectively earthed. The TRV voltage and time characteristics are described in two columns only and thus are all underdamped oscillations.

4.2 Standard TRV Derivation

Standard TRVs are derived by considering a system representation that incorporates all of the equivalent R, L and C elements and then applying circuit reduction to derive the applicable RLC circuit for analysis. This concept is illustrated in Figure 4.2, which shows a three-phase diagram for the first-pole-to-clear on an effectively earthed system.

The system consists of remote sources behind n transmission lines, local sources and a discontinuity at the end of a transmission line of length d from the circuit breaker clearing a three-phase-to-earth fault. From Section 3.4, again applying the concept of superposition—current injection with the sources shorted out—gives the sequence network shown in Figure 4.3 (refer also to Section 3.5). Simple circuit calculation shows that Z_{eq}, L_{eq} and C_{eq} are given by

$$Z_{eq} = \frac{Z_L}{n} = \frac{1}{n} \cdot \frac{3Z_1 Z_0}{2Z_0 + Z_1},$$

where Z_L is the surge impedance of a single line,

$$L_{eq} = \frac{3L_1 L_0}{2L_0 + L_1}$$

and

$$C_{eq} = \frac{1}{3}(C_0 + 2C_1).$$

For transmission lines $Z_0/Z_1 = 1.6$, giving $Z_{eq} = 1.14 Z_1/n$. For effectively earthed systems, $L_0 = 3L_1$ and $L_{eq} = 1.3 L_1$, which readers should recognize as being the expected result for the first-pole-to-clear. C_1 and C_0 are composites of all equipment capacitances at the circuit breaker location.

Table 4.2 IEC TRV requirements for 72.5 kV circuit breakers.

Rated voltage, U_r (kV)	Test duty	First-pole-to-clear factor, k_{pp} (pu)	Amplitude factor, k_{af} (pu)	TRV peak value, u_c (kV)	Time, t_3 (µs)	Time delay, t_d (µs)	Voltage, u' (kV)	Time, t' (µs)	RRRV, u_c/t_3 (kV/µs)
72.5	T100	1.5	1.4	124	165	25	41.4	80	0.75
	T60	1.5	1.5	133	73	11	44.4	35	1.82
	T30	1.5	1.6	142	36	5	47.4	18	3.94
	T10	1.5	1.7	151	36	5	50.3	18	4.19

Terminal Faults

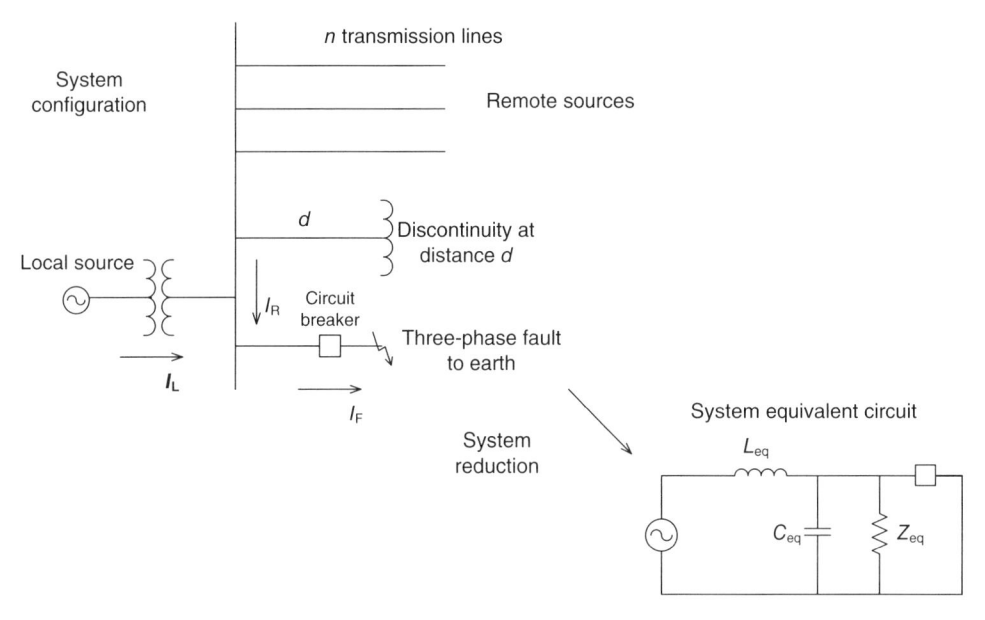

Figure 4.2 Single line diagram for first-pole-to-clear TRV calculation for an effectively earthed system.

The reduced current injection circuit is as shown in Figure 4.4, and readers will note its similarity to the circuit of Figure 2.7. Z_{eq} is resistive and is equivalent to the R value in calculations of Section 2.5. If even only one transmission line is involved, the TRV oscillation will be overdamped (aperiodic), and this permits a simpler approach that ignores C_{eq}. On this

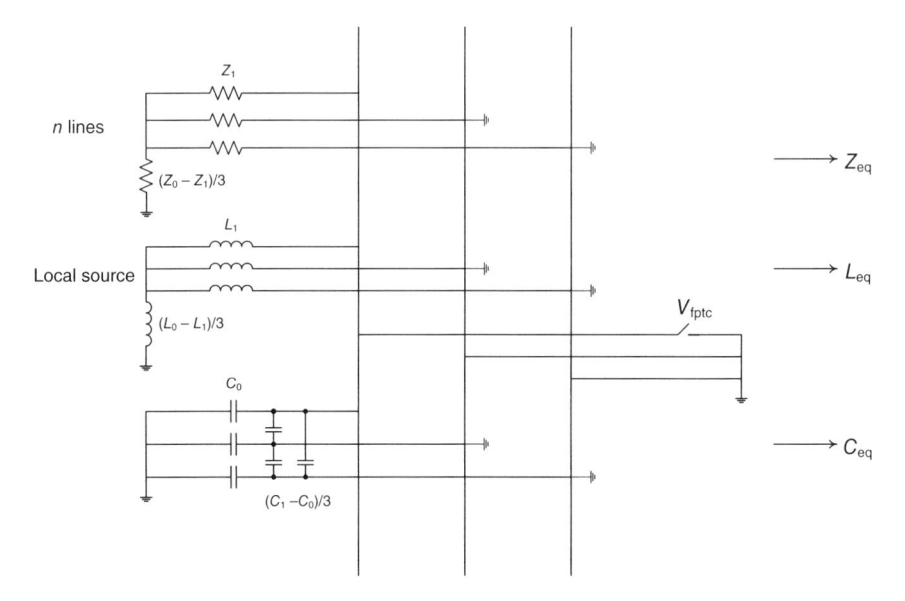

Figure 4.3 Sequence network for first-pole-to-clear calculation for an effectively earthed system.

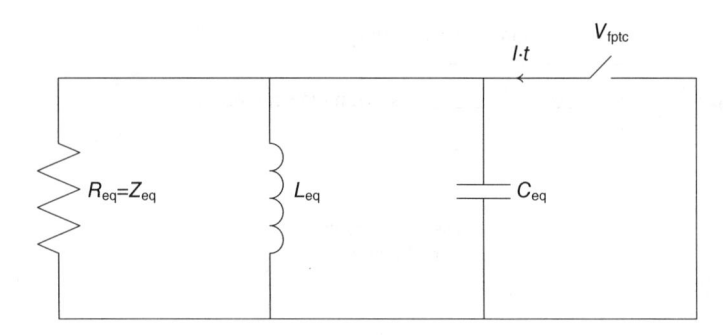

Figure 4.4 Current injection circuit for first-pole-to-clear TRV calculation.

basis, the equation for the TRV voltage v is given by

$$\frac{v}{Z_{eq}} + \frac{1}{L}\int v \cdot dt = I \cdot t.$$

Differentiating twice gives the homogeneous second-order differential equation

$$\frac{d^2v}{dt^2} + \frac{Z_{eq}}{L_{eq}}\frac{dv}{dt} = 0,$$

whose solution is

$$v = k_1 e^{r_1 t} + k_2 e^{r_2 t}.$$

From Eqs. (A.4) and (A.5), we can determine that $r_1 = 0$ and $r_2 = -Z_{eq}/L_{eq}$. The boundary conditions $v = 0$ at $t = 0$ and $v = V$ at $t = \infty$ show that $k_2 = -k_1 = V$. $V = \sqrt{2}\omega I L_{eq}$ because before the current interruption the voltage is dropped across the source impedance, I being the rms short-circuit current. v can now be written as

$$v = \sqrt{2}I\omega L_{eq}\left(1 - e^{-t/(L_{eq}/Z_{eq})}\right). \tag{4.1}$$

Differentiating Eq. (4.1) gives the rate of change of voltage RRRV:

$$\frac{dv}{dt} = \sqrt{2}\omega I Z_{eq}\, e^{-t/(L_{eq}/Z_{eq})}, \tag{4.2}$$

which at $t = 0$ becomes

$$\frac{dv}{dt} = \sqrt{2}\omega I Z_{eq}$$

or

$$\frac{dv}{dt} = Z_{eq}\frac{di}{dt} \tag{4.3}$$

because $di/dt = \sqrt{2}\omega I$.

Terminal Faults

Test duties T100 and T60 cover cases involving transmission lines. Equation (4.1) shows that the value of v is unchanged by the decrease in current T100 to T60 because L_{eq} has to increase in the same proportion. However, the rate of rise will increase because the value of Z_{eq} given by Z_L/n increases in a greater proportion than the decrease in current. These considerations are reflected in the T100 and T60 TRV values of Table 4.1.

Two points with respect to this calculation are worth further comment:

1. Ignoring C_{eq} in the calculation has little or no effect on the overall voltage calculation. However, the time delay (column 9 in Table 4.1) will be affected because $t_d = Z_{eq}C_{eq}$. The effect will be illustrated by example in Section 4.3.
2. Equation (4.3) is applicable only to overdamped-type TRVs such as those for T100 and T60.

The TRV, as represented by Eq. (4.1), has an additional consequence in that it sends a travelling wave out from the circuit breaker bus. Taking a basic single-wave approach, the wave is assumed to travel over a line with a surge impedance Z_L and encounters a discontinuity at a distance d. Taking the extreme case in which the discontinuity is an open circuit, the incident exponential wave is fully reflected and returns toward the circuit breaker bus as shown in Figure 4.5.

When the return wave reaches the circuit breaker bus, it is superimposed on the initial transient to give a double exponential voltage as shown in Figure 4.6.

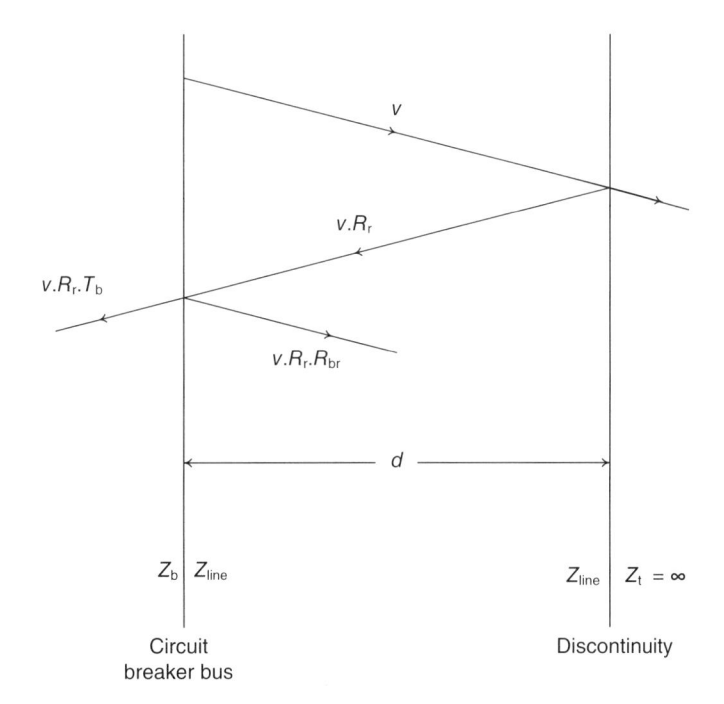

Figure 4.5 Travelling wave representation for first-pole-to-clear TRV calculation.

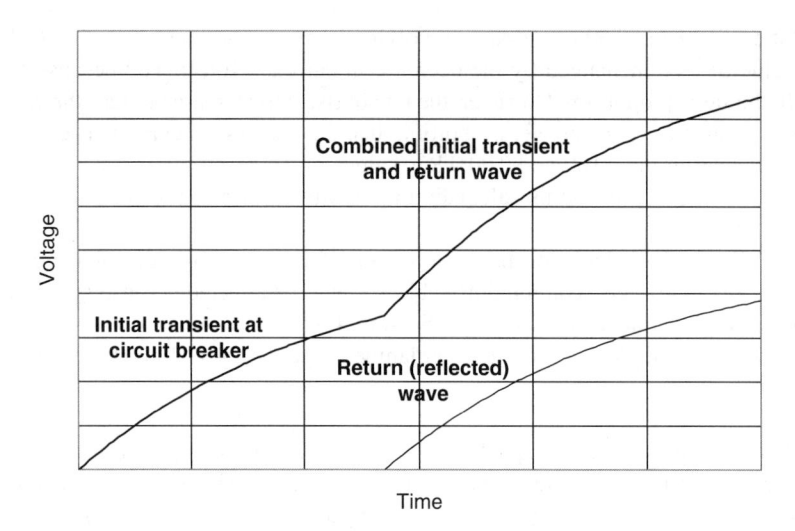

Figure 4.6 Double exponential TRV caused by initial circuit breaker bus transient and reflected wave.

The single-wave approach is a simplification. In reality, the outbound initial travelling wave follows multiple paths, reaching multiple discontinuities and generating multiple return waves. However, the single-wave approach can be considered as conservative with an additional assumption that the line termination at the circuit breaker bus consists only of L_{eq}, that is, the damping effect of Z_{eq} is not considered.

From Section 2.7, Eq. (2.52) applies for the inductive termination case, and substituting the parameters for this case, we can write

$$Z_{eq}i + L_{eq}\frac{di}{dt} = 2V\left(1 - e^{-t/(L_{eq}/Z_{eq})}\right),$$

and differentiating yields

$$\frac{d^2i}{dt^2} + \frac{Z_{eq}}{L_{eq}}\frac{di}{dt} = \frac{2V}{L_{eq}}\frac{1}{\tau}e^{-t/\tau}, \qquad (4.4)$$

where $\tau = L_{eq}/Z_{eq}$.

This is a non-homogeneous second-order differential equation, and the solution is the sum of the complementary and particular equations (Appendix A). The complementary equation is

$$\frac{d^2i_c}{dt^2} + \frac{Z_{eq}}{L_{eq}}\frac{di_c}{dt} = 0,$$

and its solution is

$$i_c = \frac{2V}{Z_{eq}}\left(1 - e^{-t/\tau}\right),$$

as readers can easily verify by applying the principles described in Appendix A.

Terminal Faults

The particular equation is

$$\frac{d^2 i_p}{dt^2} + \frac{1}{\tau}\frac{di_p}{dt} = \frac{2V}{L_{eq}}\frac{1}{\tau}e^{-t/\tau}, \qquad (4.5)$$

and the method of undetermined coefficients suggests a solution of the form (Table A.2)

$$i_p = Ct\, e^{-t/\tau},$$

where C is a constant to be determined. We can write

$$\frac{di_p}{dt} = C e^{-t/\tau}(1 - t/\tau),$$

$$\frac{d^2 i_p}{dt^2} = -C e^{-t/\tau}\left(\frac{2}{\tau} - \frac{t}{\tau^2}\right),$$

and substitution in the left-hand side of Eq. (4.5) gives

$$C e^{-t/\tau}\left[-\left(\frac{2}{\tau} - \frac{t}{\tau^2}\right) + \frac{1}{\tau}\left(1 - \frac{t}{\tau^2}\right)\right] = -C\frac{1}{\tau}e^{-t/\tau}.$$

Taking the next step and completing both sides of Eq. (4.5) gives the value of C as

$$C = -\frac{2V}{L_{eq}}.$$

The solution for Eq. (4.5) is

$$\begin{aligned} i &= i_c + i_p \\ &= \frac{2V}{L_{eq}}\left(1 - e^{-t/\tau}\right) - \frac{2V}{L_{eq}} t\, e^{-t/\tau} \\ &= \frac{2V}{L_{eq}}\left[1 - e^{-t/\tau}(1 + t/\tau)\right]. \end{aligned}$$

The additional voltage v_r contributed by the returning wave is given by

$$\begin{aligned} v_r &= L_{eq}\frac{di}{dt} \\ &= 2V\frac{t}{\tau}e^{-t/\tau}. \end{aligned}$$

v_r will reach a maximum value when $t = \tau$ and

$$\begin{aligned} v_{r\,max} &= 2V/e \\ &= 0.736 V. \end{aligned}$$

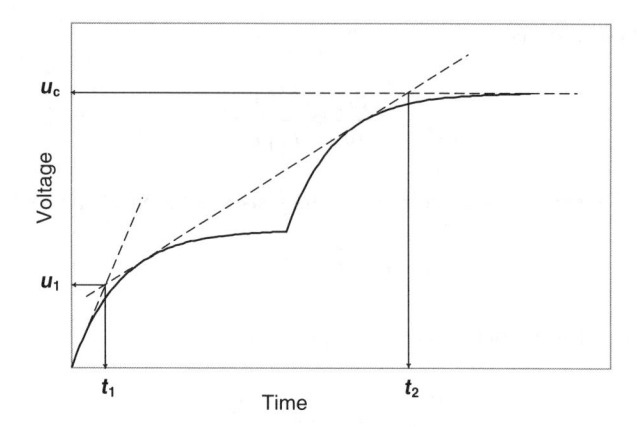

Figure 4.7 Overdamped TRV characterization.

This maximum value represents an amplitude factor of 1.736 for the overall initial plus return wave transient. However, studies show that the actual amplitude factors are lower and have values of 1.4 and 1.5 pu for T100 and T60, respectively. Note that the amplitude factor in these cases is not quite the same as that discussed for underdamped oscillations in Section 2.5 (Case 3) and is better described as a quasi-amplitude factor.

We have described the origin and nature of the overdamped system case TRVs and now need to express the result in standardized terms. As shown in Figure 4.7, the TRV is characterized by drawing a tangent to the initial transient voltage followed by a second tangent to the initial and return wave transient voltages and finally a horizontal line at the peak value of the composite TRV. The intersection point of the two tangents is defined at a voltage u_1, occurring at a time t_1. u_1 is given by

$$u_1 = 0.75 \times k_{pp} \times U_r,$$

and t_1 is determined from the required rate of rise of recovery voltage.

For the 245 kV circuit breaker and T100 case in Table 4.1,

$$u_1 = 0.75 \times 1.3 \times \frac{245\sqrt{2}}{\sqrt{3}}$$
$$= 195 \text{ kV},$$
$$t_1 = 195/2$$
$$= 98 \text{ μs}.$$

The intersection of the second tangent and the horizontal line is defined at voltage u_c occurring at time t_2. u_c is given by

$$u_c = k_{pp} \times k_{af} \times U_r,$$

and t_2 is taken as $4t_1$, representing the arrival of the standard return wave for the T100 case.

Again, for the same 245 kV circuit breaker and T100 case,

$$u_c = 1.3 \times 1.4 \times \frac{245\sqrt{2}}{\sqrt{3}}$$
$$= 364 \text{ kV}$$

and

$$t_2 = 4 \times 98$$
$$= 392 \text{ μs}.$$

A similar calculation is applicable for the T60 case. As discussed earlier, the u_1 voltage value is unchanged from that for T100 case, but the RRRV u_1/t_1 increases to 3 kV/μs. Also, the ratio t_2/t_1 increases to 6.

The time delay t_d and parameters u' and t' describe the time delay line. The ultimate prospective TRV of the test circuit (without the possible influence of the circuit breaker) is the required $u_1-t_1-u_c-t_2$ envelope without any excursion beyond the time delay line. The overall so-called four-parameter TRV representation is as shown in Figure 4.8. In the type test, the prospective TRV must meet or, better still, exceed the envelope and avoid discussions relating to measurement tolerances.

For the T30 case, the fault current is assumed to come from local sources with no involvement of transmission lines. The TRV is therefore underdamped and described by two parameters, the peak value u_c and a time t_3. t_3 is the time at which a tangent to the transient voltage intersects the horizontal u_c line, and the so-called two-parameter TRV representation is shown in Figure 4.9 with relevant time delay features.

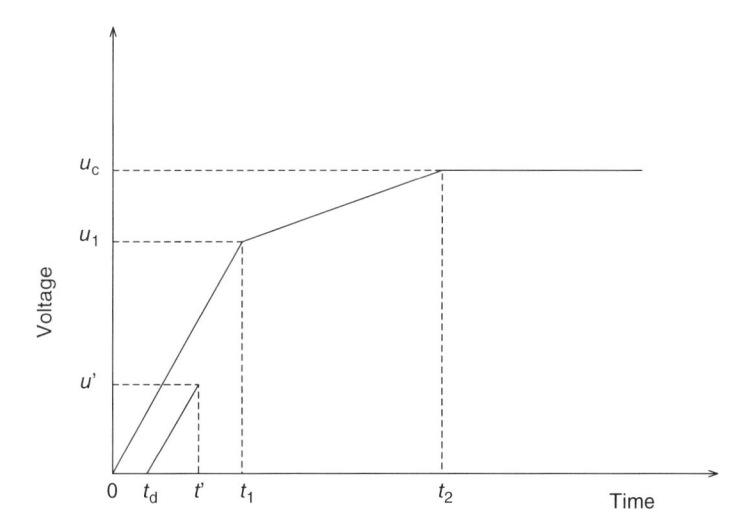

Figure 4.8 Four-parameter TRV representation.

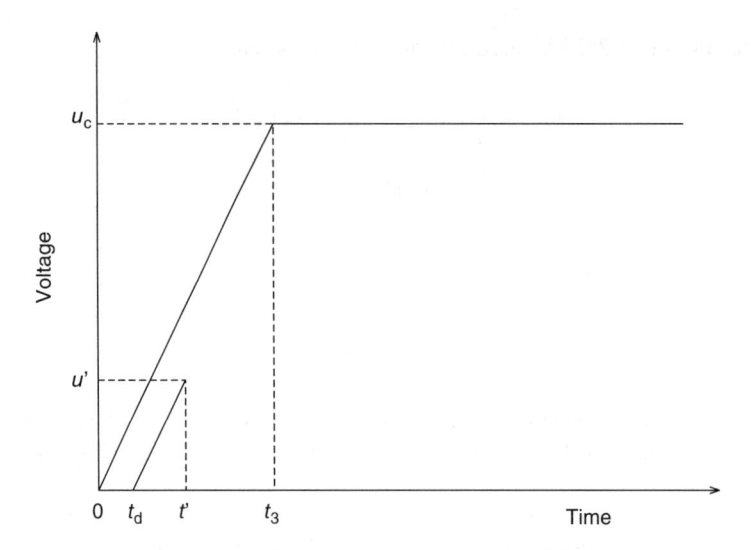

Figure 4.9 Two-parameter TRV representation.

For the 245 kV circuit breaker (Table 4.1),

$$u_c = 1.3 \times 1.54 \times \frac{245\sqrt{2}}{\sqrt{3}}$$
$$= 400 \text{ kV}$$

and $t_3 = 400/5 = 80 \, \mu s$.

The T10 case TRV is also a two-parameter one and, as explained in Section 4.1, represents a transformer-fed fault (sometimes referred to as a TFF). For the 245 kV circuit breaker,

$$u_c = 1.5 \times 0.9 \times 1.9 \times \frac{245\sqrt{2}}{\sqrt{3}}$$
$$= 459 \text{ kV}$$

and $t_3 = 459/7 = 66 \, \mu s$.

For comparison, the TRV requirements for a 72.5 kV circuit breaker are shown in Table 4.2. The system is assumed to be non-effectively earthed, and the first-pole-to-clear factors are all 1.5 pu. Also, only local fault current sources are involved (i.e. no transmission lines), and all of the TRVs are underdamped.

4.3 Effect of Added Capacitance

In Section 2.5, we derived the general equations applicable to the TRVs for the terminal fault case as summarized in Tables 2.1 and 2.2. The degree of damping is given by d_p expressed as

$$d_p = \frac{R_C}{R} = \frac{0.5\sqrt{L/C}}{R}. \tag{4.6}$$

Terminal Faults

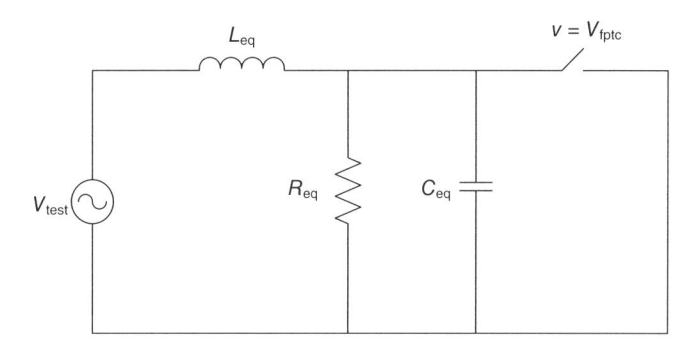

Figure 4.10 Single-phase test circuit.

To deal with excessive RRRVs, it is common practice to add capacitance; additionally, there is also the case in which shunt capacitor banks are applied in substations with existing circuit breakers. With reference to Eq. (4.6), the addition of capacitance increases the value of C, thereby decreasing the value of d_p and increasing the peak values of the TRVs. At the same time, the value of $t_g = \sqrt{LC}$ will increase, resulting in longer times to peak and lower RRRV values.

To illustrate the effect of added capacitance, we must first establish a base case of zero added capacitance. In any substation, the TRV is unknown unless it has been measured or accurately calculated. However, the TRVs to which the circuit breakers have been type tested are known, and we will use this fact to establish the base case. Starting from the TRV characteristics specified in Table 4.1 for 245 kV circuit breakers, we will calculate the first-pole-to-clear TRV equations for the four test duties T100, T60, T30 and T10. The intent is to derive a single-phase test circuit that will give the correct AC recovery voltage and the correct transient recovery voltage. The circuit to be derived is shown in Figure 4.10, and taking the case of a 245 kV, 50 kA, 60 Hz circuit breaker, we will derive the TRV equations for the above-noted fault duties using the knowledge acquired in Chapters 2 and 3 and in Section 4.2.

- T100 test duty.

  The test voltage V_{test} for the first-pole-to-clear is

$$V_{test} = 1.3 \times \frac{245}{\sqrt{3}}$$
$$= 184 \, \text{kV}.$$

1 pu test voltage peak $V_{test\ pk}$ for TRV determination is

$$V_{test\ pk} = 1.3 \times \frac{245\sqrt{2}}{\sqrt{3}}$$
$$= 260 \, \text{kV}.$$

Referring back to the generic curves of Figure 2.8, 1 pu = 260 kV, and we now have the vertical axis established.

The test current is 50 kA and L_{eq} is given by

$$L_{eq} = \frac{184}{377} \cdot \frac{1}{50}$$
$$= 0.0097 \text{ H}.$$

The value of R_{eq} is determined from Eq. (4.3).
dv/dt from Table 4.1 is $2\,\text{kV/}\mu\text{s}$ and di/dt is given by

$$\frac{di}{dt} = \sqrt{2} \times 377 \times 50 \times 10^{-6}$$
$$= 0.0266 \text{ kA/}\mu\text{s},$$
$$R_{eq} = Z_{eq}$$
$$= \frac{2}{0.0266}$$
$$= 75.2 \, \Omega.$$

Because the TRV is overdamped, C_{eq} is determined from the time delay t_d. Taking the value of $2\,\mu\text{s}$ from Table 4.1,

$$t_d = Z_{eq} C_{eq}$$
$$= R_{eq} C_{eq}$$

and

$$C_{eq} = \frac{2 \times 10^{-6}}{75.2}$$
$$= 26 \text{ nF}.$$

To derive the equation for the T100 TRV, we need to determine the values of d_p and t_g.

$$R_C = 0.5 \sqrt{\frac{L_{eq}}{C_{eq}}}$$
$$= 0.5 \sqrt{\frac{0.0097}{26 \times 10^{-9}}}$$
$$= 305 \, \Omega,$$
$$d_p = \frac{R_C}{R_{eq}}$$
$$= \frac{305}{75.2}$$
$$= 4,$$
$$\sqrt{d_p^2 - 1} = 3.87,$$
$$\frac{d_p}{\sqrt{d_p^2 - 1}} = 1.0336,$$

Terminal Faults

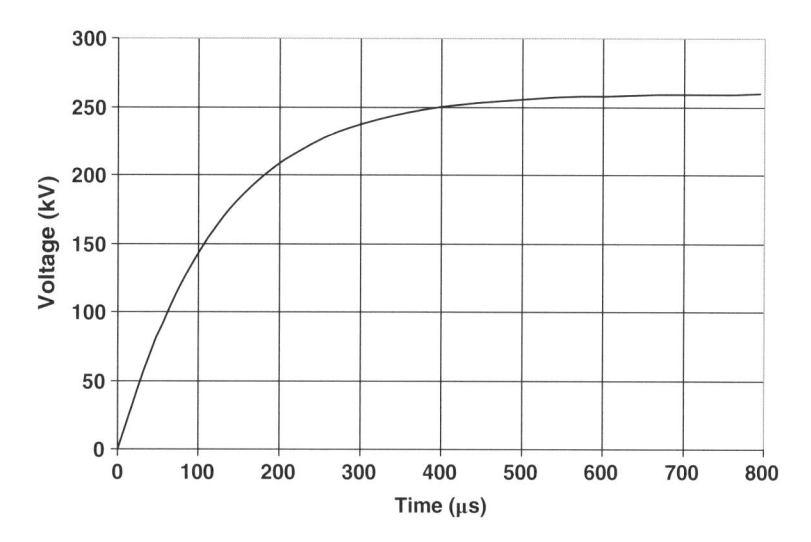

[Figure: Voltage (kV) vs Time (μs) curve rising from 0 to ~260 kV over 800 μs]

Figure 4.11 T100 TRV for a 245 kV, 50 kA, 60 Hz circuit breaker.

and from Table 2.2 the equation for the T100 TRV is

$$v = 260\left[1 - e^{-4t_g}\left(\cosh 3.87 t_g + 1.0336 \sinh 3.87 t_g\right)\right] \quad (4.7)$$

and

$$\begin{aligned} 1 \text{ pu } t_g &= \sqrt{L_{eq} C_{eq}} \\ &= \sqrt{0.0097 \times 26 \times 10^{-9}} \\ &= 15.88 \, \mu s. \end{aligned}$$

Referring again back to the generic curves of Figure 2.8, we now know the value of 1 pu t_g on the horizontal axis and that axis is now established in real time. The curves now represent a generic set of curves for the T100 case for varying values of d_p, that is, only R_{eq} is varied.

Equation (4.7) is plotted in Figure 4.11 in real time. Note that we are considering only the initial transient part of the TRV without regard to the later return wave.

As discussed earlier, because the TRV is overdamped (aperiodic), C_{eq} may be ignored, and v can be expressed in real time in the form of Eq. (4.1):

$$\begin{aligned} v &= 260\left(1 - e^{-t/(L_{eq}/R_{eq})}\right) \\ &= 260\left(1 - e^{7752t}\right), \end{aligned} \quad (4.8)$$

where t is in seconds.

Equations (4.7) and (4.8) are plotted in Figure 4.12, and it is clearly evident that they are essentially equal. However, there is a difference in the initial part of the TRV in that the

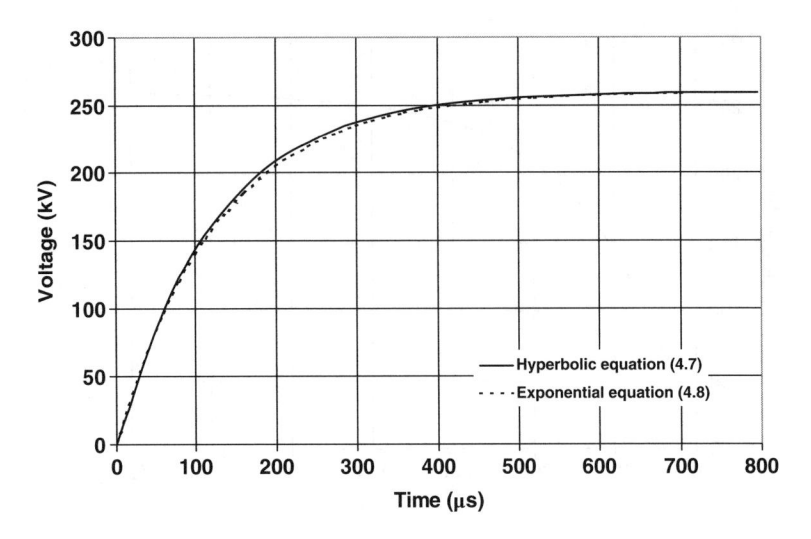

Figure 4.12 T100 TRV hyperbolic versus exponential representation for a 245 kV, 50 kA, 60 Hz circuit breaker.

exponential representation does not show the time delay component at all as illustrated in Figure 4.13.
- T60 test duty.
 The equation for this test duty can be derived in an identical manner as for T100 and using the appropriate values for the test current and from Table 4.1. Readers can readily determine that

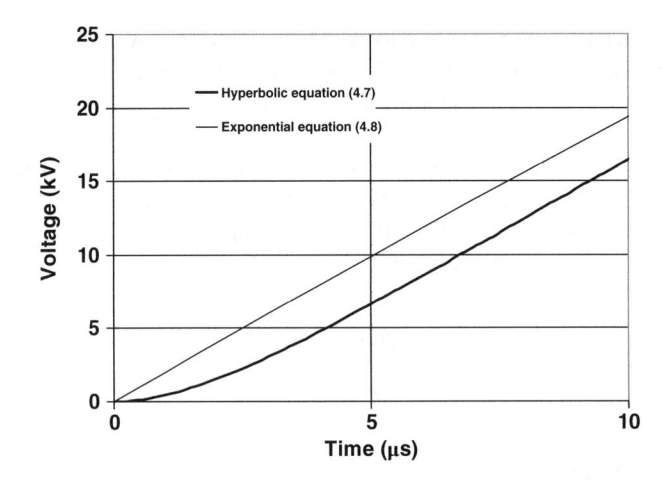

Figure 4.13 T100 TRV time delay region hyperbolic versus exponential representation for a 245 kV, 50 kA, 60 Hz circuit breaker.

Terminal Faults

$$V_{test} = 184 \text{ kV},$$
$$V_{test\ pk} = 260 \text{ kV},$$
$$L_{eq} = 0.016 \text{ H},$$
$$R_{eq} = 187.5 \text{ }\Omega,$$
$$C_{eq} = 10.6 \text{ nF},$$
$$R_C = 614 \text{ }\Omega,$$
$$d_p = 3.27,$$
$$1 \text{ pu } t_g = 13 \text{ }\mu s,$$

and the equation for the TRV is

$$v = 260\left[1 - e^{-3.27\ t_g}\left(\cosh 3.1 t_g + 1.05 \sinh 3.1 t_g\right)\right]. \tag{4.9}$$

Equation (4.9) is plotted in Figure 4.14 in real time.
- T30 test duty.
  For this case, the AC components are derived as for the T100 and T60 test duties:

$$V_{test} = 184 \text{ kV},$$
$$V_{test\ pk} = 260 \text{ kV},$$
$$L = 0.0325 \text{ H}.$$

However, the TRV is now underdamped, and from Table 4.1, $k_{af} = 1.54$, leading directly to the value for d_p. From Eq. (2.46) and Figure 2.11, $R/R_C = 5.2$ and $d_p = 0.192$.

The remaining unknown quantity is C_{eq} and is determined from its relationship to L_{eq} and not from the time delay as for the overdamped test duties. The time to peak value of the TRV T is given by

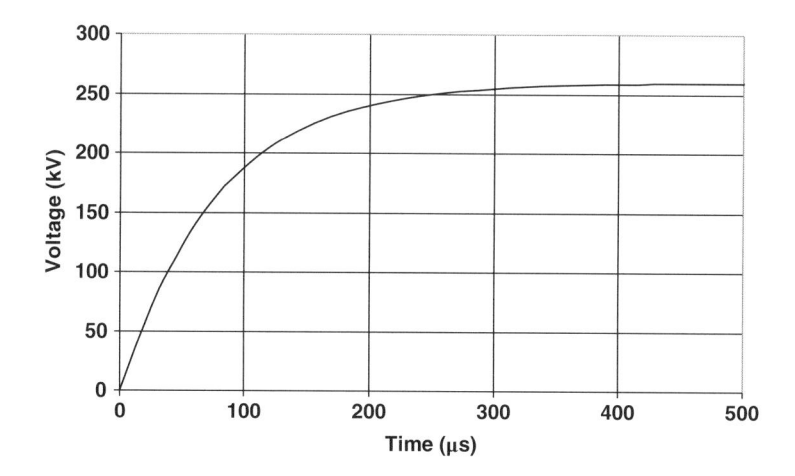

$$T = \pi\sqrt{L_{eq} C_{eq}}. \tag{4.10}$$

Figure 4.14 T60 TRV for a 245 kV, 50 kA, 60 Hz circuit breaker.

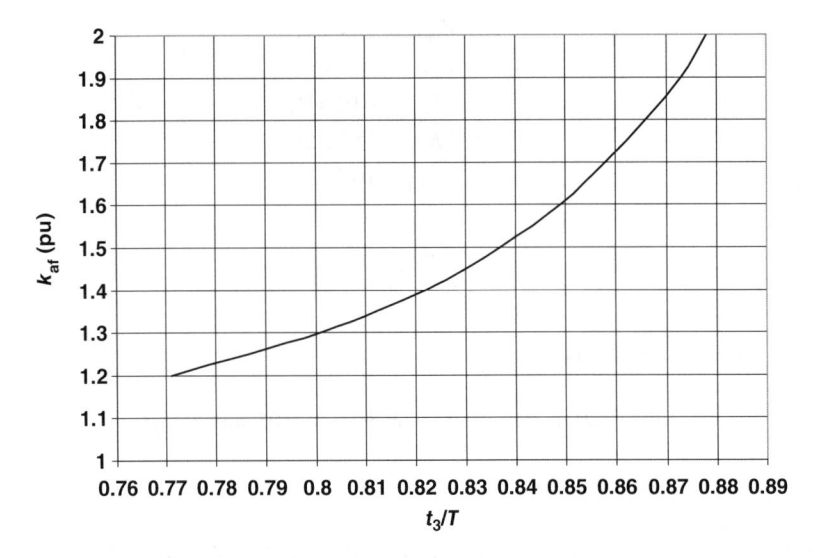

Figure 4.15 Ratio t_3/T for underdamped oscillations.

Table 4.1 gives the value of t_3 to be 80 μs, and we need to determine the relationship t_3/T for $k_{af} = 1.54$ pu. The relationship can be determined in general by graphical means, that is, using Figure 2.8 as a basis and Excel, draw the tangents for the varying underdamped oscillations and the intersection of the tangents with a horizontal line through the peaks gives the applicable t_3 values. t_3/T values as a function of k_{af} are shown in Figure 4.15.

From Figure 4.15, $t_3/T = 0.842$,

$$T = \frac{80}{0.842} = 95 \text{ μs},$$

and it follows from Eq. (4.10) that $C_{eq} = 28$ nF.

$$R_C = 0.5\sqrt{\frac{0.0325}{28 \times 10^{-9}}}$$
$$= 538 \text{ Ω},$$
$$R_{eq} = 5.2 \times 538$$
$$= 2797 \text{ Ω},$$
$$d_p = 0.192,$$
$$\sqrt{1 - d_p^2} = 0.98,$$
$$\frac{d_p}{\sqrt{1 - d_p^2}} = 0.196,$$
$$1 \text{ pu } t_g = \sqrt{0.0325 \times 28 \times 10^{-9}}$$
$$= 30 \text{ μs}.$$

Terminal Faults

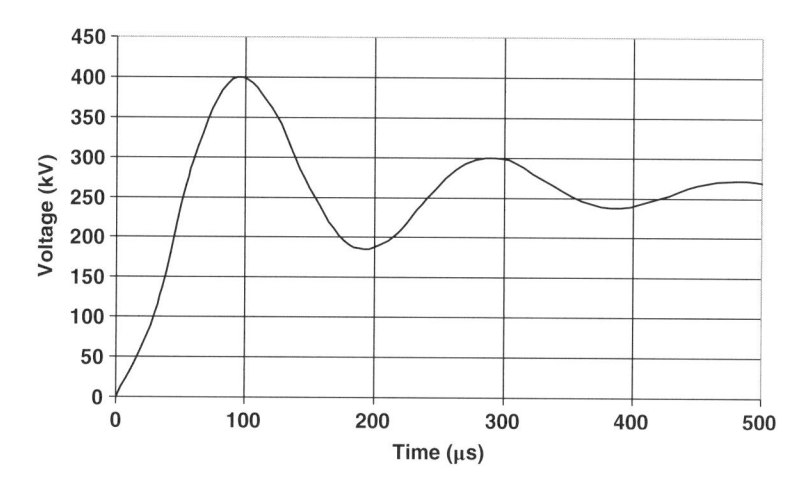

Figure 4.16 T30 TRV for a 245 kV, 50 kA, 60 Hz circuit breaker.

From Table 2.2, the equation for the T30 test duty is

$$v = 260\left[1 - e^{-0.192 t_g}\left(\cos 0.98 t_g + 0.196 \sin 0.98 t_g\right)\right]. \tag{4.11}$$

Equation (4.11) is plotted in Figure 4.16.
- T10 test duty.

The equation for this test duty is derived in an identical manner as for the T30 test duty, noting that the first-pole-to-clear k_{pp} factor is now 1.5 pu. Readers can verify that the applicable test circuit parameters are

$$V_{test} = 212 \text{ kV},$$
$$V_{test\ pk} = 300 \text{ kV},$$
$$L_{eq} = 0.112 \text{ H},$$
$$C_{eq} = 5 \text{ nF},$$
$$R_C = 2366 \text{ }\Omega,$$
$$R_{eq} = 11832 \text{ }\Omega,$$
$$d_p = 0.2,$$
$$1 \text{ pu } t_g = 23.7 \text{ }\mu s.$$

The equation for the TRV is

$$v = 300\left[1 - e^{-0.2 t_g}\left(\cos 0.98 t_g + 0.2 \sin 0.98 t_g\right)\right]. \tag{4.12}$$

Equation (4.12) is plotted in Figure 4.17, and the TRVs for all four test duties are shown in Figure 4.18.

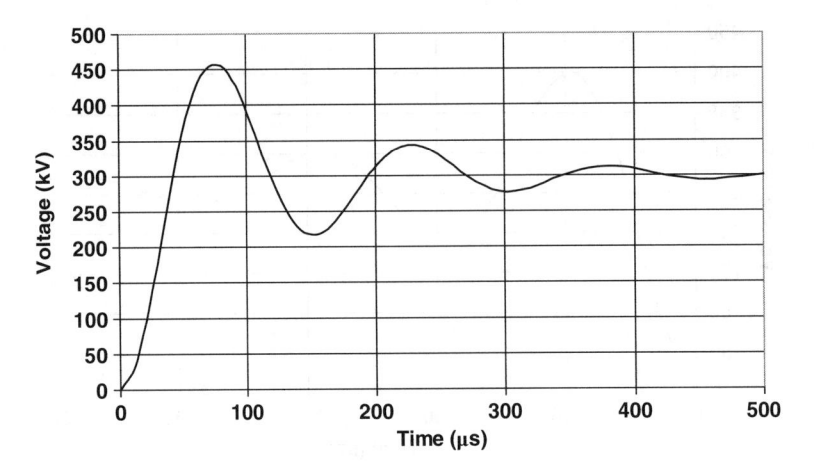

Figure 4.17 T10 TRV for a 245 kV, 50 kA, 60 Hz circuit breaker.

To now consider the effect of added capacitance C_{add}, this capacitance appears in parallel with C_{eq} as shown in Figure 4.19. The AC components are, of course, unchanged as are L_{eq} and R_{eq}. The effective C_{eq} increases in turn, affecting R_C, d_p and t_g. For the T100 test duty taking $C_{add} = 1\,\mu F$, the revised quantities are

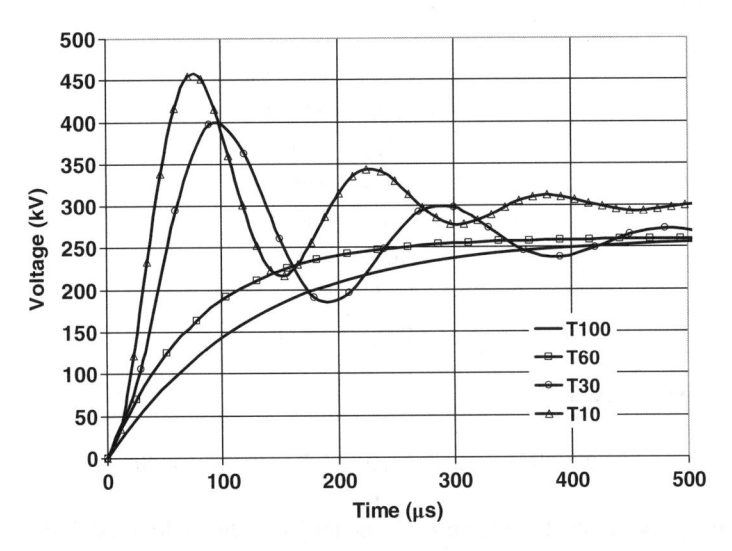

$$C_{eq} = 1.026\text{ nF},$$
$$R_C = 48.6\,\Omega,$$
$$d_p = 0.64,$$
$$1\text{ pu }t_g = 99.7\,\mu s.$$

Figure 4.18 T100, T60, T30 and T10 TRVs for a 245 kV, 50 kA, 60 Hz circuit breaker.

Terminal Faults

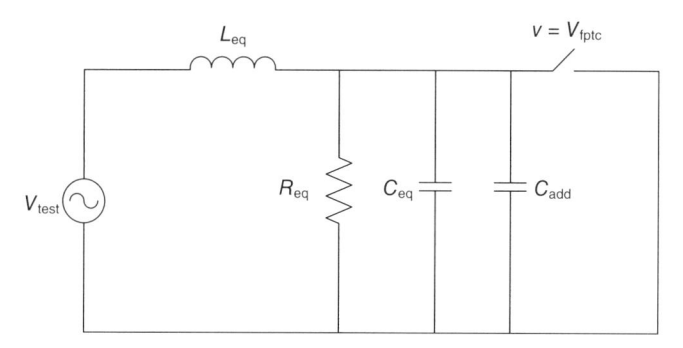

Figure 4.19 Single-phase test circuit with added capacitance.

d_p is less than 1, and the circuit has become underdamped; the equation for the TRV is

$$v = 260\left[1 - e^{-0.64t_g}\left(\cos 0.76t_g + 0.84 \sin 0.76t_g\right)\right]. \tag{4.13}$$

Equation (4.7), T100 TRV with no added capacitance, and Eq. (4.13) are plotted in Figure 4.20. The equations for all four test duties with an added capacitance of 1 μF are given in Table 4.3. The individual plots for T60, T30 and T10 without and with 1 μF added capacitance are shown in Figures 4.21–4.23, respectively. The combined overall effect for all four test duties is shown in Figure 4.24. Added capacitance clearly drives the TRVs in the direction of underdamping: upward to increase the peak value and to the right to decrease the RRRV value.

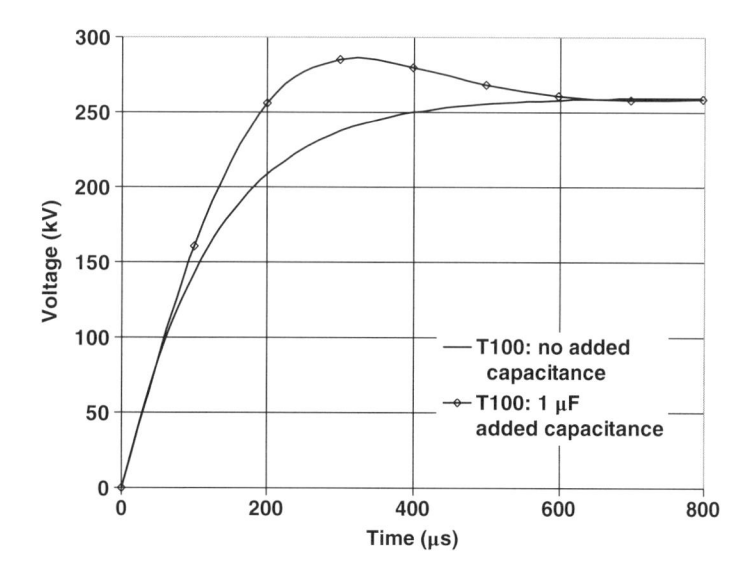

Figure 4.20 T100 TRV for a 245 kV, 50 kA, 60 Hz circuit breaker without and with 1 μF added capacitance.

Table 4.3 Equations for T100, T60, T30 and T10 test duties with 1 μF added capacitance.

Test duty	Equation	1 pu t_g (μs)
T100	$v = 260\left[1 - e^{-0.64 t_g}\left(\cos 0.76 t_g + 0.84 \sin 0.76 t_g\right)\right]$	99.7
T60	$v = 260\left[1 - e^{-0.335 t_g}\left(\cos 0.94 t_g + 0.35 \sin 0.94 t_g\right)\right]$	127
T30	$v = 260\left[1 - e^{-0.03 t_g}\left(\cos t_g + 0.03 \sin t_g\right)\right]$	182
T10	$v = 300\left[1 - e^{-0.014 t_g}\left(\cos t_g + 0.014 \sin t_g\right)\right]$	335

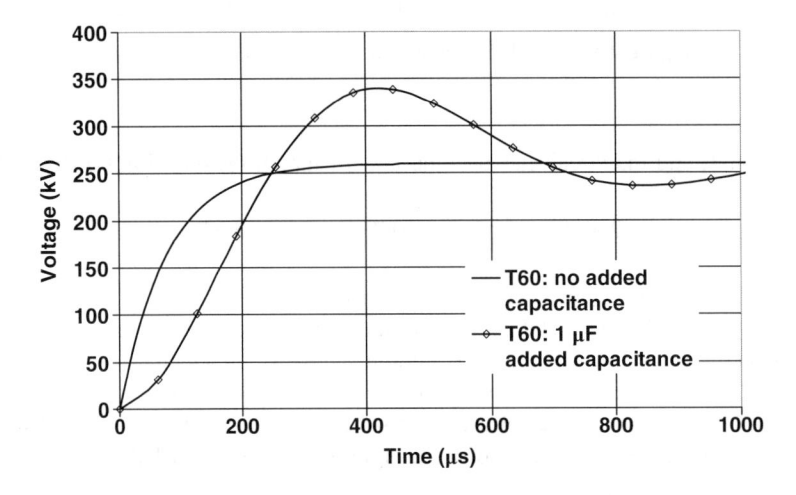

Figure 4.21 T60 TRV for a 245 kV, 50 kA, 60 Hz circuit breaker without and with 1 μF added capacitance.

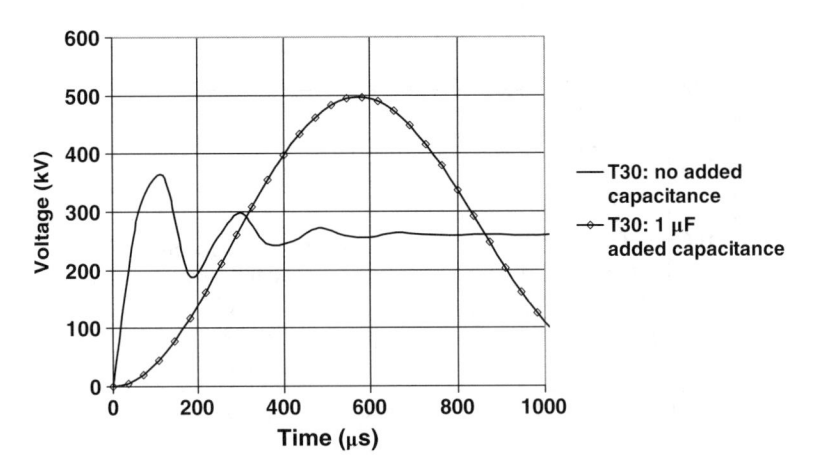

Figure 4.22 T30 TRV for a 245 kV, 50 kA, 60 Hz circuit breaker without and with 1 μF added capacitance.

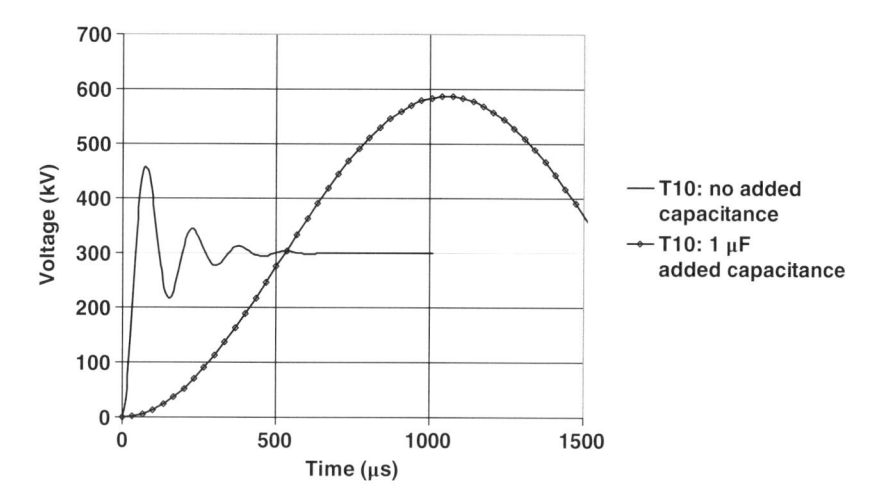

Figure 4.23 T10 TRV for a 245 kV, 50 kA, 60 Hz circuit breaker without and with 1 μF added capacitance.

Important note: Readers should not overreact to the possible higher TRV peak values. In reality, actual substation TRVs will be less than those used in above calculations. Simulation studies with proper source representations are required to determine if any issues need to be addressed for existing circuits following the addition of shunt capacitor banks.

4.4 Effect of Added Resistance

Added resistance is achieved in reality by means of opening resistors. Opening resistors were widely used with air-blast circuit breakers but fell into disuse with the development of SF_6 gas circuit breakers having superior interrupting capability. However, with the advent of ultrahigh-voltage (UHV) transmission at 1000 kV and above, opening resistors are again used. The economic and technical advantages of opening resistors lie in being able to use fewer interrupters in series per circuit breaker pole.

IEC TRV requirements for an 1100 kV circuit breaker are shown in Table 4.4, and we will use these requirements to demonstrate the effect of opening resistors assuming a short-circuit rating of 50 kA at 50 Hz. The requirements are obviously noteworthy for their high values but also for the first-pole-to-clear factor k_{pp} being 1.2 pu for all four test duties. As shown in Figure 4.25, the opening resistor appears in parallel with R_{eq}, reducing its effective value. Viewed in theory, R_C and t_g are not affected, but d_p will increase (according to Eq. (4.6)), driving the TRVs in the direction of overdamping.

The base TRV test circuit parameters are derived in exactly the same way as those in Section 4.3 and are listed in Table 4.5. The corresponding equations are given in Table 4.6 and plotted in Figure 4.26.

800 Ω is a typical opening resistor value, and the addition of this resistor to the base circuits will revise the parameters as given in Table 4.7. The corresponding equations are listed in Table 4.8. Judging solely from the changes in the equations from Table 4.6 to Table 4.8, the effect of the added resistance is minimal for T100 and T60 and significant for T30 and T10. The TRV for T30 becomes critically damped and T10 overdamped.

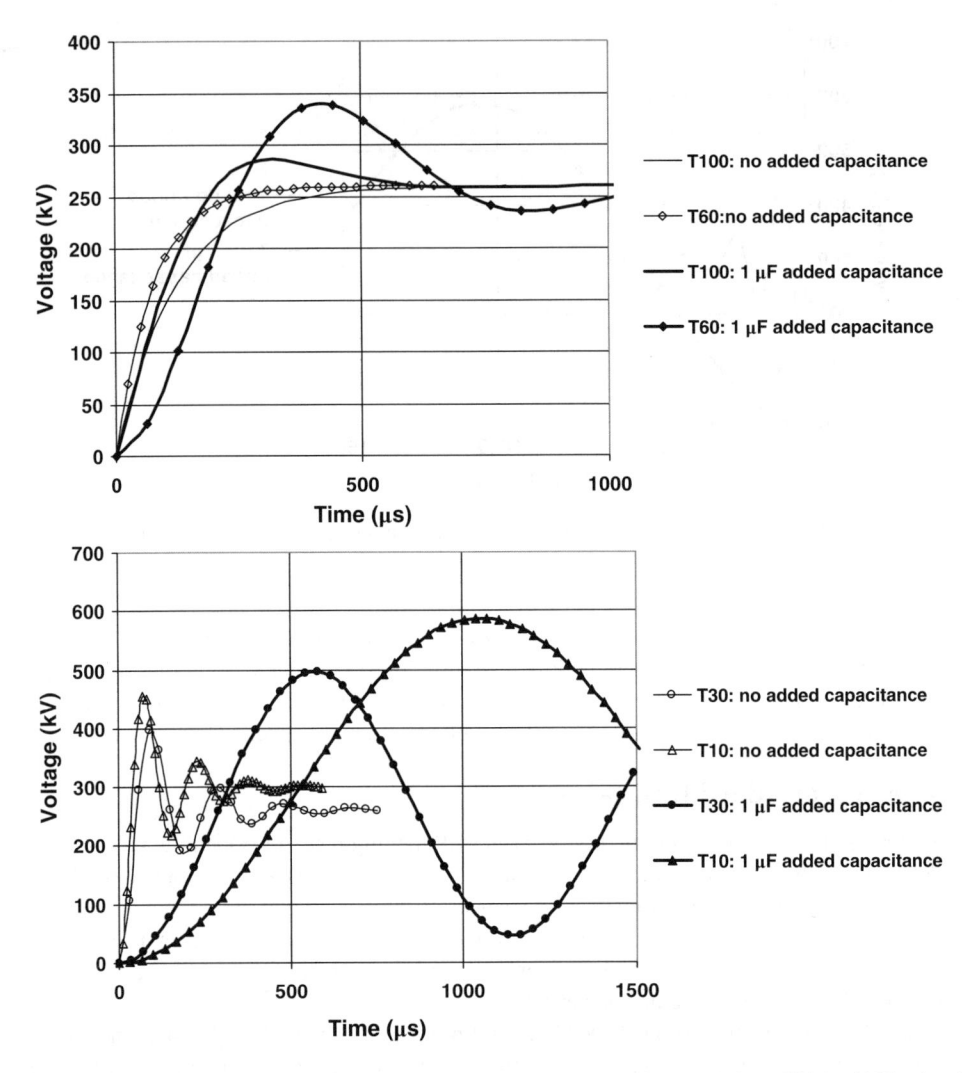

Figure 4.24 Overall effect of 1 μF added capacitance on TRVs for a 245 kV, 50 kA, 60 Hz circuit breaker.

4.5 Effect of Added Inductance

Added inductance in practice is the addition of series reactors for the purpose of load sharing or to limit fault current magnitudes. The effect of series reactors cannot be viewed simply by reference to Eq. (4.6) as for the added capacitance and resistance cases. The reason for this is that the series reactor also adds its inherent capacitance to the circuit, and the TRV circuit representation is as shown in Figure 4.27. An immediate conclusion is that the circuit is no longer single frequency but rather double or even multiple frequency. In addition, because the value of C_R is in the order of hundreds of picofarads, the frequency of the $L_R C_R$ component is high, producing a TRV that may exceed the capability of the circuit breakers in the vicinity of

Table 4.4 IEC TRV requirements for 1100 kV circuit breakers.

Rated voltage, U_r (kV)	Test duty	First-pole-to-clear factor, k_{pp} (pu)	Amplitude factor, k_{af} (pu)	First reference voltage, u_1 (kV)	Time, t_1 (μs)	TRV peak value, u_c (kV)	Time, t_2 or t_3 (μs)	Time delay, t_d (μs)	Voltage, u' (kV)	Time, t' (μs)	Rate of rise, u_1/t_1, u_c/t_3 (kV/μs)
1100	T100	1.2	1.50	808	404	1617	1212	2 (113)	404	204 (315)	2
	T60	1.2	1.50	808	269	1617	1212	2–81	404	137–216	3
	T30	1.2	1.54	—	—	1660	332	50	553	161	5
	T10	1.2	1.76	—	—	1897	271	41	632	131	7

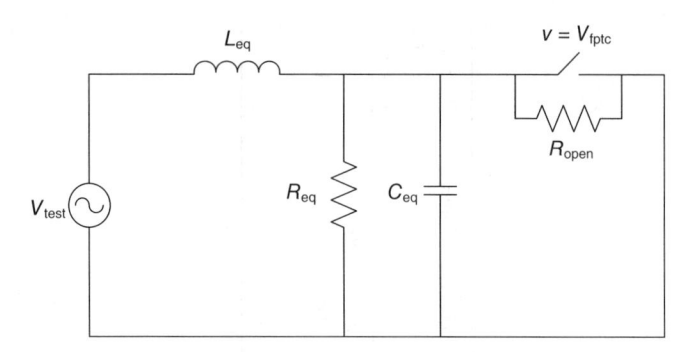

Figure 4.25 Test circuit with added opening resistor.

Table 4.5 1100 kV circuit breaker test circuit parameters for T100, T60, T30 and T10 at 50 Hz.

Test duty	V_{test}	$V_{test\ pk}$	R_{eq} (Ω)	L_{eq} (H)	C_{eq} (nF)	R_C (Ω)	d_p
T100	762	1078	90.9	0.0485	22	742	8.16
T60			225	0.081	8	1591	7.07
T30			3322	0.16	98	639	0.192
T10			27 082	0.485	20	2462	0.09

the series reactors. To mitigate this transient, it is common practice to add a line to earth capacitance as part of the series reactor installation.

The circuit of Figure 4.27 does not lend itself to simple analysis because of its multiple frequency nature. However, as a first approximation, we can treat the source and series reactor circuit as being independent of one another. The TRV calculation is now similar to those discussed already and will be illustrated using a real series reactor application on the French transmission system [8].

A 10 Ω series reactor was applied on the 225 kV, 50 Hz system and has the added effect of reducing the fault level from 31.5 to 9.2 kA. Source details are unknown, but we will assume that the circuit breaker is rated at 50 kA and take the source TRV as being equivalent to the T60 case because 31.5 kA is approximately 60% of the fault current rating.

The pu equation for the source TRV follows from Eq. (4.9):

$$v_{S\ pu} = 1 - e^{-3.27 t_{g1}} \left(\cosh 3.1 t_{g1} + 1.05 \sinh 3.1 t_{g1} \right), \quad (4.14)$$

where 1 pu $t_{g1} = 13$ μs and the source inductance L_s is 0.016 H.

Table 4.6 Base TRV equations for 1100 kV circuit breaker for T100, T60, T30 and T10 at 50 Hz.

Test duty	Equation	1 pu t_g (μs)
T100	$v = 1078(1 - e^{-1874t})$	—
T60	$v = 1078(1 - e^{-2778t})$	—
T30	$v = 1078\left[1 - e^{-0.192 t_g}(\cos 0.98 t_g + 0.196 \sin 0.98 t_g)\right]$	125
T10	$v = 1078\left[1 - e^{-0.09 t_g}(\cos t_g + 0.09 \sin t_g)\right]$	98.5

Terminal Faults

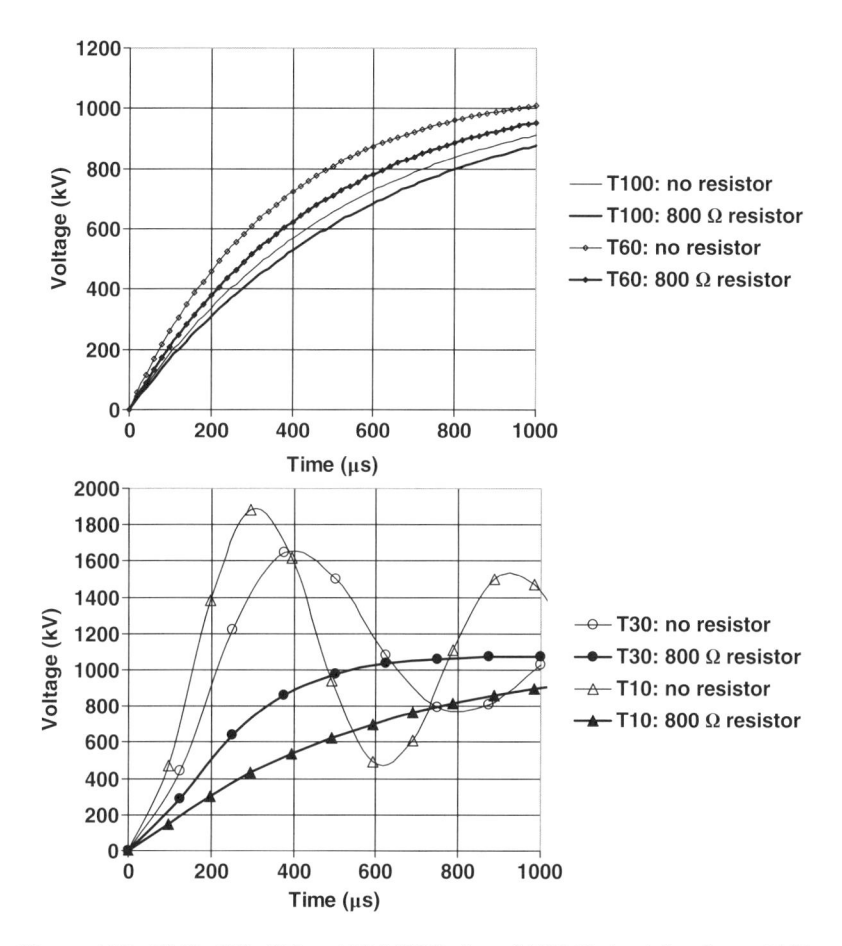

Figure 4.26 T100, T60, T30 and T10 TRVs for a 1100 kV circuit breaker at 50 Hz.

Table 4.7 1100 kV circuit breaker test circuit parameters for T100, T60, T30 and T10 with 800 Ω opening resistor.

Test duty	V_{test}	$V_{test\,pk}$	R_{eq} (Ω)	L_{eq} (H)	C_{eq} (nF)	R_C (Ω)	d_p
T100	762	1078	81.6	0.0485	22	742	9.1
T60			175	0.081	8	1591	9.1
T30			633	0.16	98	639	1
T10			775	0.485	20	2462	3.18

Table 4.8 TRV equations for 1100 kV circuit breaker for T100, T60, T30 and T10 at 50 Hz with 800 Ω opening resistor.

Test duty	Equation	1 pu t_g (μs)
T100	$v = 1078(1 - e^{-1683t})$	—
T60	$v = 1078(1 - e^{-2160t})$	—
T30	$v = 1078[1 - e^{-t_g}(1 + t_g)]$	125
T10	$v = 1078[1 - e^{-3.18 t_g}(\cosh 3t_g + 1.06 \sinh 3t_g)]$	98.5

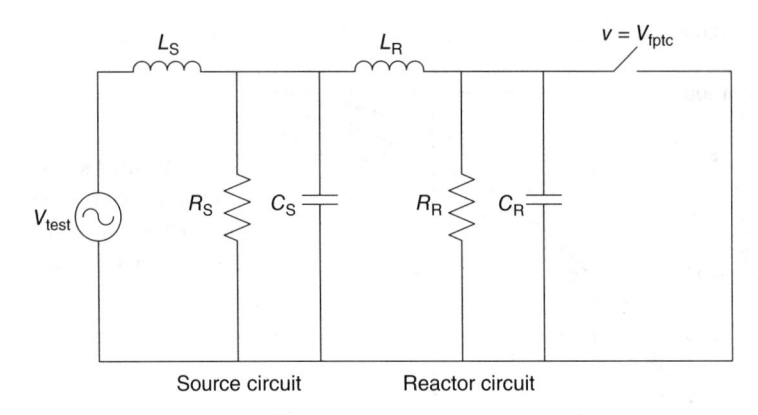

Figure 4.27 TRV circuit representation with added series reactor.

For the series reactor circuit, we will assume that $k_{af} = 1.9$ pu (a reasonable assumption because series reactors are low-loss devices) and that a capacitance of 12 nF is added to earth.

$$L_R = 1.3 \times \frac{10}{314}$$
$$= 0.0414 \text{ H},$$
$$C_R = 12 \text{ nF},$$
$$1 \text{ pu } t_{g2} = 22.29 \text{ μs}.$$

From $k_{af} = 1.9$ and Eq. (2.46) (Figure 2.11),

$$d_p = 0.033,$$

and the pu series reactor equation is

$$v_{R\,pu} = 1 - e^{-0.033 t_{g2}} \left(\cos t_{g2} + 0.033 \sin t_{g2} \right). \tag{4.15}$$

The overall TRV will be distributed between the two circuits in ratios of L_S and L_R to $L_S + L_R$, respectively, giving the combined TRV equation (based on a system voltage of 225 kV) from Eqs. (4.14) and (4.15) as

$$\begin{aligned} v_{SR} = &\, 66.4 \left[1 - e^{-3.27 t_{g1}} \left(\cosh 3.1 t_{g1} + 1.05 \sinh 3.1 t_{g1} \right) \right] \\ &+ 172.6 \left[1 - e^{-0.033 t_{g2}} \left(\cos t_{g2} + 0.033 \sin t_{g2} \right) \right]. \end{aligned} \tag{4.16}$$

The individual part and combined equations are shown in Figure 4.28 compared with the IEC T60 and T30 test duty TRVs for a 245 kV circuit breaker. Refer also to Section 7.2.1, where the same calculation is done with the series reactor on the load side of the circuit breaker.

The representation of the series reactor circuit is accurate, and the combined TRV is very dependent on the source representation. Therefore, for estimation purposes, the approximation method with a conservative choice source is recommended. Simulation software is commonly

Terminal Faults

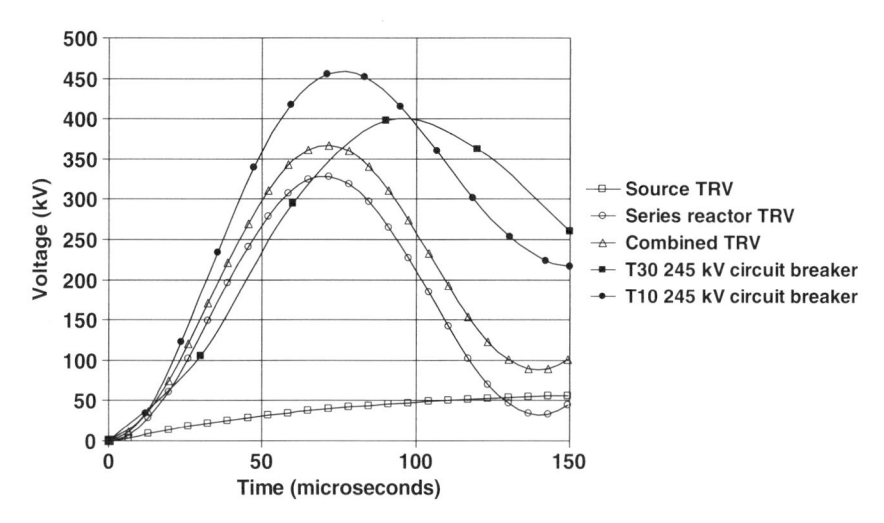

Figure 4.28 TRVs with added series reactor and underdamped source TRV representation.

used in practice for decision making with respect to existing circuit breakers and it is equally important that an accurate source representation is used.

To mention a historical approach, some readers will be aware of the method proposed by Boehne [5] to deal with certain coupled circuits. The method was treated further by Hammarlund [6] and promoted by Greenwood [9]. The method originates from the work of Johnson [4] of Bell Telephone, who studied telephone circuits and showed that certain coupled circuits could be converted to simpler non-coupled circuits having exactly the same impedance independent of frequency. Boehne surmised that Johnson's conversion could be applied to circuits of the type shown to the left in Figure 4.29 to derive a decoupled circuit as shown to the right in the figure giving a method of dealing with TRVs associated with double frequency circuits. The conversion is possible only if resistance is ignored and the TRV should

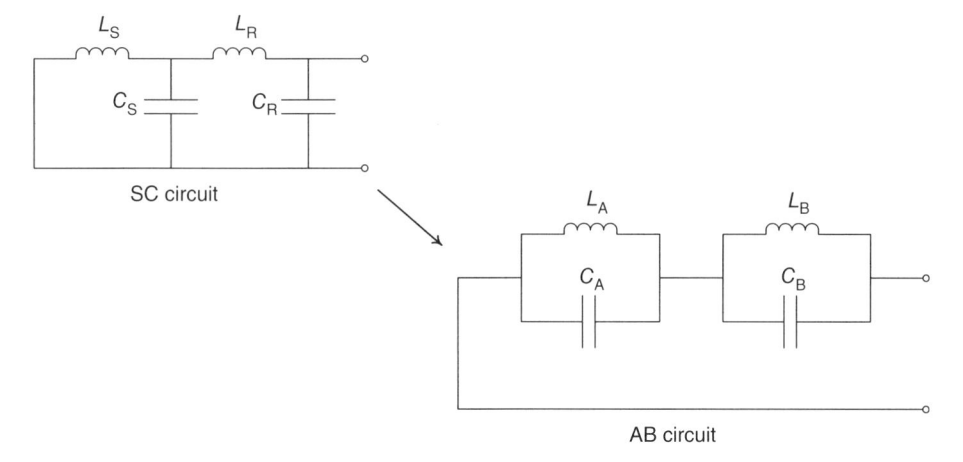

Figure 4.29 Johnson's conversion for coupled to decoupled circuits.

yield a conservative result. The solution for the TRV will take the form of the sum of two $(1 - \cos)$ functions. From Table 2.2 for the underdamped case, the equation for the TRV in per unit is

$$v = 1 - e^{-d_p t_g}\left(\cos\sqrt{1 - d_p^2}\, t_g + \frac{d_p}{\sqrt{1 - d_p^2}}\sin\sqrt{1 - d_p^2}\, t_g\right). \qquad (4.17)$$

For $R = \infty$, $d_p = 0$ and Eq. (4.17) becomes

$$v = 1 - e^{-(0)t_g}\left(\cos t_g + (0)\sin t_g\right)$$
$$= 1 - \cos t_g$$
$$= 1 - \cos \omega t \quad \text{since } t_g = \omega t,$$

and the TRV for the converted AB circuit has the form

$$v_{AB} = a_A(1 - \cos \omega_A t) + a_B(1 - \cos \omega_B t), \qquad (4.18)$$

where

$$a_A = \frac{L_A}{L_A + L_B}$$

and

$$a_B = \frac{L_B}{L_A + L_B}$$
$$= 1 - a_A.$$

The conversion calculation is convoluted,[1] and Boehne provided nomograms for the purpose. The nomograms are difficult to use, and Hammarlund provided an easier to use stepwise calculation method. The method assumes zero damping and cannot be used for the above example. It is difficult not to view it with some scepticism, and its use is not recommended in principle and for reasons discussed in Section 7.2.1.2.

4.6 Out-of-Phase Switching

Out-of-phase switching is a variation of the terminal fault case and occurs when a circuit breaker connecting two power systems (or parts of the same system) is inadvertently closed when the systems are not synchronized. The basic circuit for the out-of-phase switching condition is shown in Figure 4.30. The magnitude of the current is standardized at 25% of the circuit breaker rated short-circuit current to reflect the sum of the two source impedances and additional transmission line impedances. The standardized TRV has the lowest RRRV and the

---

[1] Neither Boehne nor Hammarlund provided details justifying the conversion method for transient calculations.

Terminal Faults

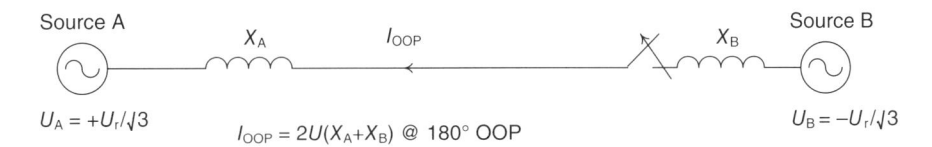

Source A X_A I_{OOP} X_B Source B

$U_A = +U_r/\sqrt{3}$ $I_{OOP} = 2U/(X_A+X_B)$ @ 180° OOP $U_B = -U_r/\sqrt{3}$

Figure 4.30 Basic circuit for out-of-phase switching condition.

highest peak value for the terminal fault cases as shown in Figure 4.31 for a 245 kV circuit breaker at rated out-of-phase switching current. The TRV is a four-parameter one and therefore overdamped with a first-pole-to-clear factor k_{pp} equal to 2 pu reflecting a worst case 180° phase difference between the two systems.

At current interruption, each system will oscillate independently, and the TRV seen by the circuit breaker will be the difference between the two individual system TRVs. The latter TRVs are of opposite polarities and, with reference to Figure 2.8 and no transmission line, the starting point of each TRV oscillation is $\sqrt{2}I(X_A - X_B)$, I being the rated out-of-phase switching current. This situation is illustrated in Figure 4.32, where the following can be noted: the axis of the oscillations will be plus or minus 1 pu of the individual system voltage peaks, thus scaling the vertical axis; however, the per-unit t_g values are unlikely to be the same for both oscillations, meaning that the two oscillations have to be real time in order to calculate the difference correctly and derive the TRV across the circuit breaker. The axis of oscillation of this TRV will be the 2 pu system voltage peak.

4.7 Asymmetrical Currents

As discussed in Section 1.3 and shown in Figures 1.1–1.3, circuit breakers have to be capable of interrupting both symmetrical and asymmetrical currents. From a TRV perspective, interrupting the latter currents is less onerous than that for the former currents because the current zero crossings are shifted away from the recovery voltage peaks to a lower point on the

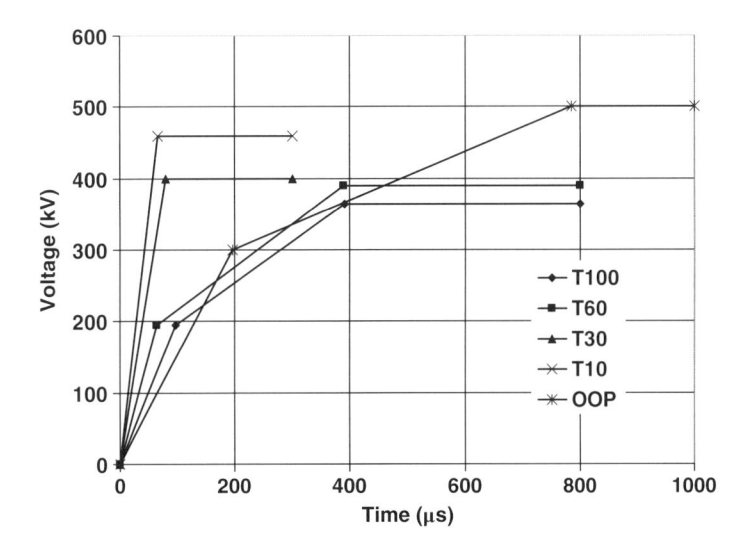

Figure 4.31 Out-of-phase switching TRV compared with other terminal fault TRVs.

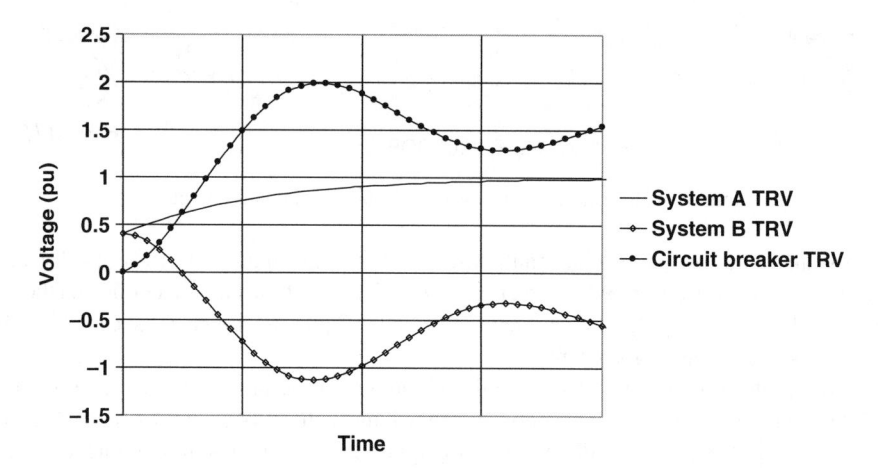

Figure 4.32 Illustration of out-of-phase switching individual system TRVs.

voltage half-cycle. This is clear from Figure 4.33; although the circuit amplitude factor is, of course, unaffected, the axis of oscillation has shifted to a lower value, resulting in a less onerous TRV. In contrast, the circuit breaker will be thermally more stressed because of the major current loops compared with the symmetrical case.

Asymmetrical currents are calculated using the circuit shown in Figure 4.34. In this case, the fault current i is initiated by closing the circuit breaker and applying a sinusoidal voltage v to the RL circuit, where R and L are those of the system. We can write

$$v = V_{pk} \cos(\omega t + \alpha),$$

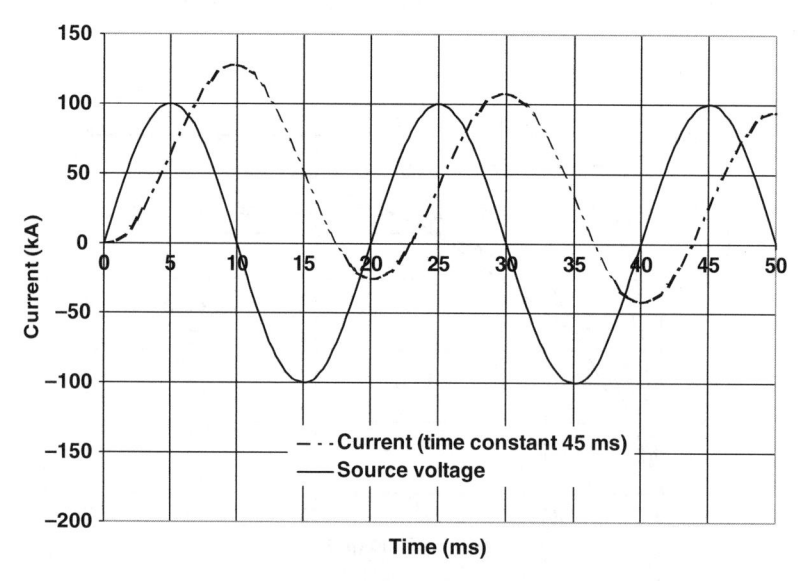

Figure 4.33 Asymmetrical current in relation to the recovery voltage.

Terminal Faults

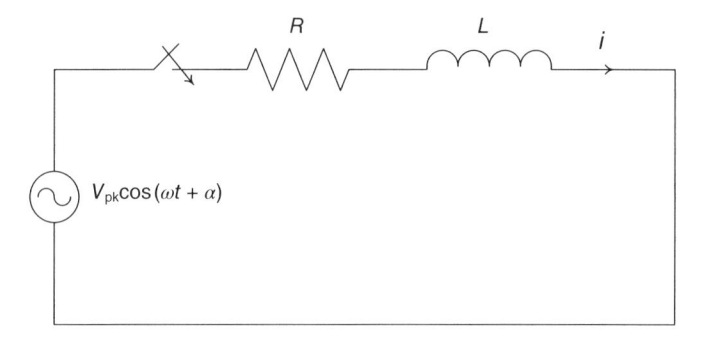

[Figure: Circuit diagram showing AC source $V_{pk}\cos(\omega t + \alpha)$ connected through a switch to resistor R and inductor L in series, with current i.]

Figure 4.34 Circuit for asymmetrical current calculation.

where α is closing angle relative to the voltage half-cycle, for example, for $\alpha = 0$, closing at time zero means closing at the voltage peak and for $\alpha = 90°$ at time zero will result in closing at the voltage zero crossing. Applying Kirchoff's voltage law,

$$L\frac{di}{dt} + Ri = V_{pk}\cos(\omega t + \alpha),$$

and differentiating across gives

$$\frac{d^2 i}{dt^2} + \frac{R}{L}\frac{di}{dt} = -\frac{V\omega}{L}\sin(\omega t + \alpha). \quad (4.19)$$

The solution is of the form

$$i = i_1 + i_2,$$

where i_1 in turn is the solution of the complementary equation

$$\frac{di^2}{dt^2} + \frac{R}{L}\frac{di}{dt} = 0 \quad (4.20)$$

and i_2 is the solution for the particular equation

$$\frac{di_2^2}{dt^2} + \frac{R}{L}\frac{di^2}{dt} = -\frac{V\omega}{L}\sin(\omega t + \alpha). \quad (4.21)$$

The solution for Eq. (4.20) is of the form

$$i_1 = A_1 e^{r_1 t} + A_2 e^{r_2 t},$$

which reduces to

$$i_1 = A_1 + A_2 e^{-t/(L/R)} \quad (r_1 = 0)$$

and further to

$$i_1 = A_2\, e^{-t/(L/R)} \quad (i_1 = 0 \text{ in steady state}).$$

The solution for Eq. (4.21) is (Table A.2)

$$i_2 = A\cos(\omega t + \alpha) + B\sin(\omega t + \alpha),$$

and from Eqs. (A.22) and (A.23),

$$-aA\omega^2 + bB\omega + Ac = 0,$$
$$-aB\omega^2 - BA\omega + cB = -\frac{V\omega}{L}.$$

Substituting $a = 1$, $b = R/L$ and $c = 0$ gives

$$-A\omega^2 + \frac{R}{L}B\omega = 0,$$
$$-B\omega^2 - \frac{R}{L}A\omega = -\frac{V\omega}{L},$$

and solving for A and B:

$$A = \frac{V_{pk}R}{Z^2},$$
$$B = \frac{V_{pk}\omega L}{Z^2},$$

where $Z = \sqrt{R^2 + (\omega L)^2}$.

$$\begin{aligned}
i_2 &= \frac{V_{pk}}{Z}\left[\frac{R}{Z}\cos(\omega t + \alpha) + \frac{\omega L}{Z}\sin(\omega t + \alpha)\right] \\
&= \frac{V_{pk}}{Z}[\cos(\omega t + \alpha)\cos\theta + \sin(\omega t + \alpha)\sin\theta] \\
&= \frac{V_{pk}}{Z}\cos(\omega t + \alpha - \theta),
\end{aligned}$$

θ being the phase angle between the voltage and the current and is very close to 90° in this case (see Table 4.10).

Because we are dealing with short-circuit current, R is negligible compared with $\omega L = X_L$ ($\theta \approx 90°$):

$$\begin{aligned}
i_2 &= \frac{V_{pk}}{X_L}\cos(\omega t + \alpha - \theta) \\
&= \sqrt{2}I\cos(\omega t + \alpha - \theta),
\end{aligned}$$

Terminal Faults

where I is the circuit breaker rated short-circuit current.

$$i = i_1 + i_2$$
$$= \sqrt{2}I \cos(\omega t + \alpha - \theta) + A_2 e^{-t/(L/R)}.$$

At $t = 0$, $i = 0$,

$$0 = \sqrt{2}I \cos(\alpha - \theta) + A_2,$$

and the complete solution for i is

$$i = \sqrt{2}I \left[\cos(\omega t + \alpha - \theta) - e^{-t/(L/R)} \cos(\alpha - \theta) \right]. \quad (4.22)$$

Equation (4.22) has two components: the first term within the brackets gives the steady-state AC current and the second term the transient offsetting DC current.

Two extreme cases can be considered.

1. *Zero transient component*: The transient component will be zero when the $\cos(\alpha - \theta)$ term is zero. This will occur when $\alpha - \theta = \pi/2$ or $\alpha = \theta + \pi/2$, which corresponds to closing the circuit breaker at a source voltage peak and prospective AC steady-state current zero crossing.
2. *Maximum transient component*: The maximum transient component will occur when $\cos(\alpha - \theta) = 1$, that is, when $\alpha - \theta = 0$ or $\alpha = \theta$. This event corresponds to closing the circuit breaker at a source voltage zero crossing and prospective AC steady-state current peak value. The equation for the maximum current i_{max} is

$$i_{max} = \sqrt{2}I \left(\cos \omega t - e^{-t/(L/R)} \right). \quad (4.23)$$

To provide an explanation for the occurrence of asymmetrical currents, Eqs. (4.22) and (4.23) are both mathematical and electrical solutions for Eq. (4.19). Interestingly, the former solution shows the existence of a transient component under certain conditions.

Figure 4.35 shows the relationship between the source voltage and prospective AC current at 50 Hz and a phase angle $\theta = 90°$. Clearly, closing the circuit breaker at a voltage peak will result in a steady-state AC current only. In contrast, closing at a source voltage zero crossing means that the prospective AC current is at its peak value. This current cannot appear out of nowhere, that is, it must start from zero. To achieve the zero current start, the prospective AC peak current is cancelled by a DC current of equal value and opposite polarity as shown in Figure 4.35. An AC system cannot support a DC current, and that current must decay to zero with the combined AC prospective and DC currents giving the asymmetrical current as calculated from Eq. (4.23) and shown in Figure 4.35.

The DC component current cannot be taken as an injected current because it is not possible to inject a step current into an inductive circuit. Rather, it can be viewed as being by a $\sqrt{2}I$ current source that is immediately disconnected leaving a "stranded" DC current of $\sqrt{2}I$ to decay to zero through the RL series circuit. Taking this notion and applying considerations similar to those used in Chapter 2, we can now treat Eq. (4.20) as representing the current in a source-free series RL circuit with the inductance "precharged" to $\sqrt{2}I$, this being the current value at $t = 0$. On closing the circuit, the current decays to zero at the L/R time constant, which readers can readily verify.

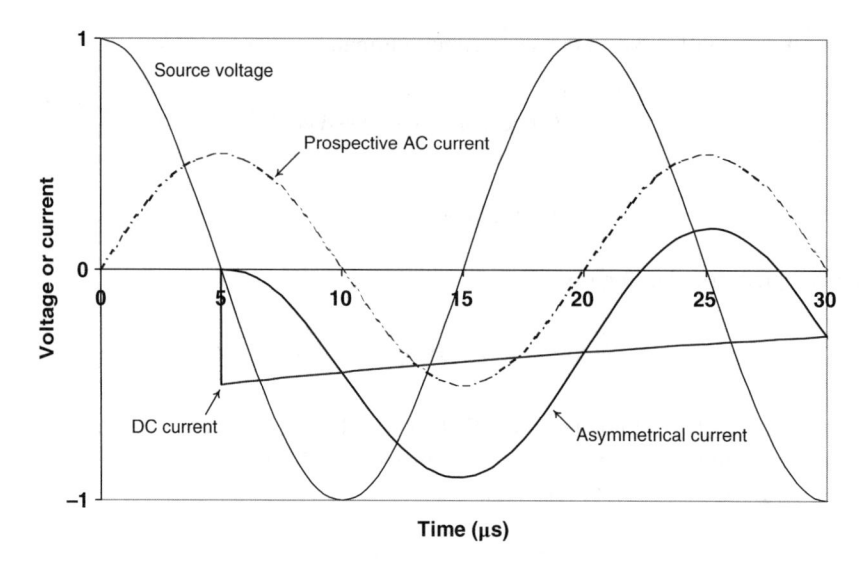

Figure 4.35 Asymmetrical current derivation.

Equation (4.23) will have a maximum at $\omega t = \pi$, and the peak value I_{pk} of the combined AC and DC components in general is given by

$$I_{pk} = \sqrt{2}I\left(1 + e^{-t/\tau}\right), \tag{4.24}$$

where τ is the time constant L/R. IEC 62271-100 allows for four possible time constants generally increasing with system voltage. These time constants are 45, 60, 75 and 120 ms.

Equation (4.24) is frequency dependent because π radians correspond to different half-cycle times at 50 and 60 Hz. For example, at $\tau = 45$ ms and considering the I_{pk}/I factor,

$$50\,\text{Hz}: \quad \sqrt{2}\left(1 + e^{-10/45}\right) = 2.547,$$

$$60\,\text{Hz}: \quad \sqrt{2}\left(1 + e^{-8.33/45}\right) = 2.589.$$

Table 4.9 gives the I_{pk}/I factors for the above-noted time constants with the standardized values shown in brackets. Figure 4.36 shows the asymmetrical current for a 63 kA, 50 Hz

Table 4.9 Ratio of I_{pk}/I for IEC 62271-100 time constants.

τ (ms)	I_{pk}/I	
	50 Hz	60 Hz
45	2.547 (2.5)	2.589 (2.6)
60	2.611 (2.7)	2.645 (2.7)
75	2.652 (2.7)	2.679 (2.7)
100	2.715 (2.7)	2.733 (2.7)

Terminal Faults

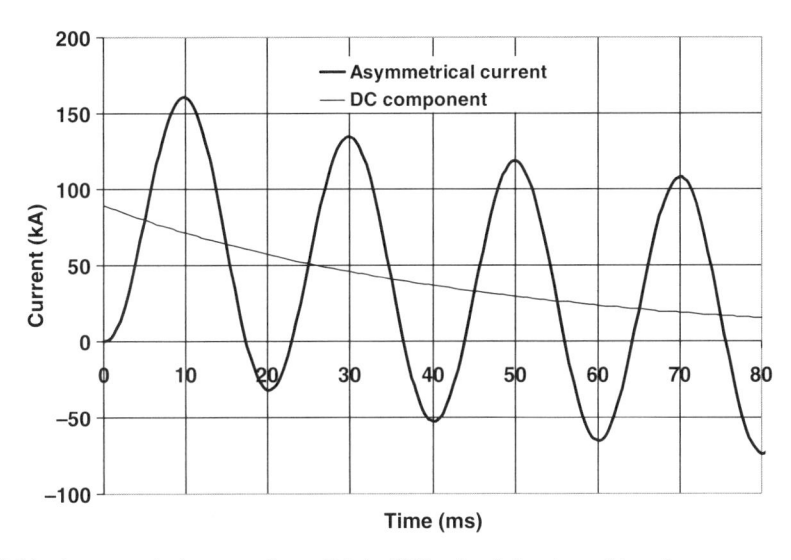

Figure 4.36 Asymmetrical current for a 63 kA, 50 Hz circuit breaker with a time constant of 45 ms.

circuit breaker with a time constant of 45 ms. In testing, the required DC component is expressed as a percentage of the AC component at contact separation.

Asymmetry is also treated by considering the X/R ratio of the system. The relationship between X/R and τ is given by

$$\frac{X}{R} = \frac{\omega L}{R} = \omega \tau,$$

and we can also write

$$\theta = \tan^{-1}\frac{\omega L}{R}$$
$$= \tan^{-1}\frac{X}{R}.$$

Table 4.10 gives the X/R and θ values for the IEC time constants.

Asymmetrical current type testing is normally done at the most common system time constant of 45 ms. However, system time constants will change with time, and the question

Table 4.10 Relationships between the IEC time constants and X/R ratios.

τ (ms)	50 Hz		60 Hz	
	X/R	θ	X/R	θ
45	14	85.9	17	86.6
60	19	86.9	22.6	87.4
75	23.5	87.5	28.3	87.9
120	37.7	88.5	45.2	88.7

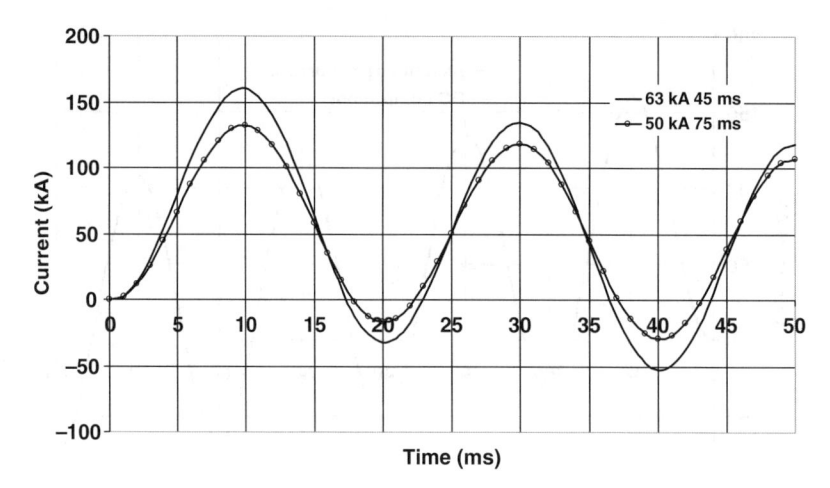

Figure 4.37 Asymmetrical currents for currents of 63 and 50 kA with time constants of 45 and 75 ms, respectively.

arises with respect to the capability of existing circuit breakers at longer time constants. This situation is treated by taking the area of the major loop before current interruption as basis and then calculating the allowable current at the longer time constant that gives the same area.

By way of example, we will take the case of the 63 kA, 45 ms circuit breaker of Figure 4.35 and determine if it can be used at 50 kA and 75 ms without resorting to a new type test. Both cases are plotted in Figure 4.37, and assuming current interruption at the current zeros between 35 and 40 ms, the last major loops are expanded in Figure 4.38.

IEC 62271-100 allows the area comparison to be made by taking the peak value of the current times the duration of the loops. On this basis, 63 kA/45 ms and 50 kA/75 ms give areas of 1.836 and 1.753 kA s, respectively, and no further type testing is required for the latter case.

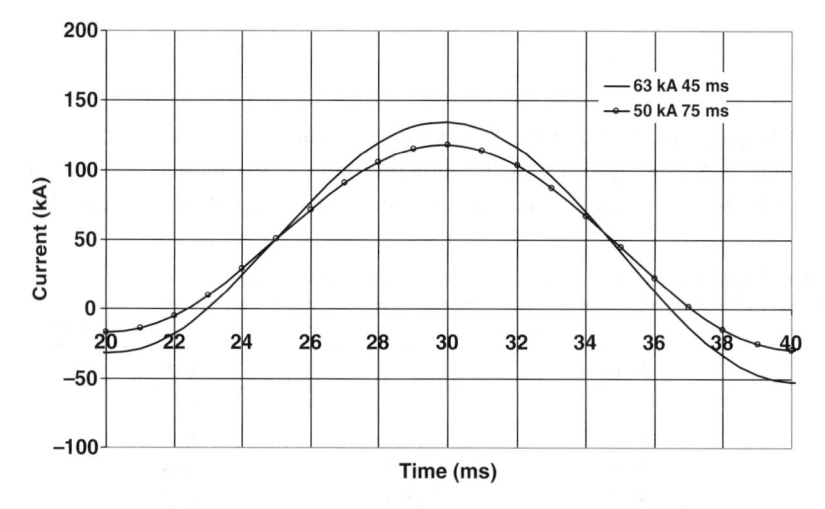

Figure 4.38 Expanded 20–40 ms range for the currents shown in Figure 4.37.

Terminal Faults

More exact calculation of the areas can be done using Excel by applying the trapezoidal rule or Simpson's rule; this approach is discussed in Appendix E.

4.8 Double Earth Faults

The double earth fault is an exceptional case and can occur on non-effectively earthed systems. The sequence of events is as follows: a fault to earth occurs on one phase on one side of a circuit breaker, for example, the load side; the fault causes the voltage to rise on the other two phases to a line voltage to earth, and a second fault to earth occurs on one of those phases but on the other side of the circuit breaker. Effectively, there is a phase-to-phase fault through earth as shown in Figure 4.39. The fault current is single phase to be interrupted by the c-pole of the circuit with a recovery voltage equal to the line voltage caused by the sustained fault on a-phase, that is, the pole factor k_{pp} equals $\sqrt{3}$.

The TRVs for this case can be either two- or four-parameter TRVs and are calculated in the same manner as for the T100 cases using the 1.732 pu pole factor and the applicable amplitude factors. The fault current is 0.866 times the fault current for the T100 symmetrical fault case.

The above quantities—k_{pp} and the fault current—are, of course, calculated using symmetrical components. The k_{pp} factor is calculated from the fault circuit and sequence network in Figure 4.40 and that for the fault current as shown in Figure 4.41.

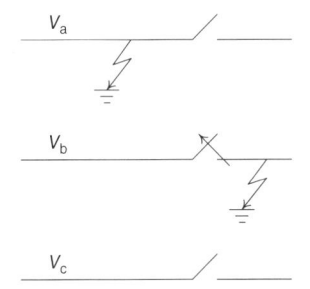

Figure 4.39 Double earth fault circuit representation.

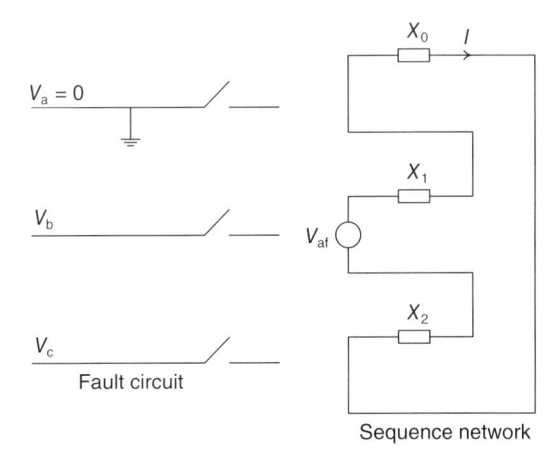

Figure 4.40 Double earth fault k_{pp} calculation.

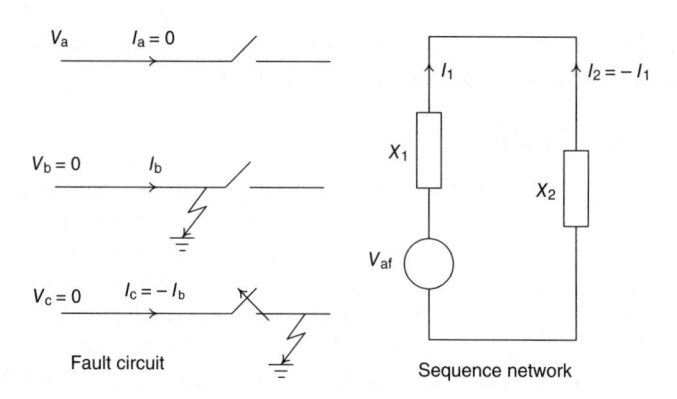

Figure 4.41 Double earth fault current calculation.

From Figure 4.39 and Eq. (3.3), we can write

$$V_0 = \frac{1}{3}(V_b + V_c),$$
$$V_1 = \frac{1}{3}(aV_b + a^2V_c),$$
$$V_2 = \frac{1}{3}(a^2V_b + aV_c),$$
$$V_0 + V_1 + V_2 = 0,$$

implying a series connection of the sequence networks (Figure 4.39). However, the current I equals zero because the phase currents I_a, I_b and I_c are all zero, and we must resort to continuing the calculation of V_b and V_c and then allow the current to go to zero.

$$I = \frac{V_{af}}{X_0 + X_1 + X_2}, \qquad (4.25)$$

$$V_0 = -IX_0,$$
$$V_1 = -IX_1 + V_{af},$$
$$V_2 = -IX_2.$$

Applying the reverse transformation (Eq. (3.4)),

$$V_a = V_0 + V_1 + V_0 = 0,$$
$$V_b = -IX_0 + a^2(-IX_1 + V_{af}) + a(-IX_2),$$
$$V_c = -IX_0 + a(-IX_1 + V_{af}) + a^2(-IX_2).$$

Substituting for I from Eq. (4.25) gives V_b and V_c as

$$V_b = V_{af}\left[a^2 - \frac{X_0 + a^2X_1 + aX_1}{X_0 + X_1 + X_2}\right], \qquad (4.26)$$

Terminal Faults

$$V_c = V_{af}\left[a - \frac{X_0 + aX_1 + a^2X_1}{X_0 + X_1 + X_2}\right]. \qquad (4.27)$$

Readers will recognize that Eqs. (4.26) and (4.27) are identical to Eqs. (3.24) and (3.25), respectively, for the single-phase-to-earth fault case in an effectively earthed system. To force the fault current to zero in that case and equally in this case (Eq. (4.25)), only X_0 can approach infinity because X_1 and X_2 cannot be infinite.

As $X_0 \to \infty$, Eqs. (4.26) and (4.27) become

$$V_b = V_{af}\left[a^2 - 1\right],$$
$$V_c = V_{af}[a - 1],$$

and taking the absolute values

$$k_{pp} = \left|\frac{V_b}{V_{af}}\right| = \left|\frac{V_c}{V_{af}}\right| = \sqrt{3}.$$

For the fault current calculation, from Figure 4.40 and Eqs. (3.3) and (3.5), respectively, we can write

$$V_0 = V_1 = V_2 = \frac{1}{3}V_a,$$
$$I_0 = 0,$$
$$I_1 = \frac{1}{3}(a - a^2)I_b,$$
$$I_2 = \frac{1}{3}(a^2 - a)I_b$$
$$= -I_1,$$
$$I_1 = \frac{V_{af}}{X_1 + X_2}$$
$$= \frac{V_{af}}{2X_1}.$$

Taking the reverse transformation (Eq. (3.6)),

$$I_a = 0,$$
$$I_b = (a^2 - a)I_1,$$
$$I_c = (a - a^2)I_1,$$
$$I_c = j\sqrt{3}\frac{V_{af}}{2}$$
$$= j0.866\frac{V_{af}}{X_1},$$
$$|I_c| = 0.866 I_{T100},$$

where I_{T100} is the fault current for the three-phase fault case (refer to Section 3.1).

4.9 Summary

Circuit breaker TRVs are fundamentally based on the oscillatory circuit and unbalanced AC circuit theory discussed in Chapters 2 and 3, respectively. For the terminal fault case, the TRVs are either overdamped or underdamped and are described by standardized first-pole-to-clear, peak and RRRV values. As discussed in Chapter 1, the values have evolved mostly in the second half of the twentieth century and are subject to continuing review as knowledge and circuit breaker technology advances.

Circuit breakers are required to have a capability to interrupt fault currents over a range from the 100% short-circuit current rating down to 10% of that rating. The capability is demonstrated by four test duties T100, T60, T30 and T10. Power systems rated at 100 kV and below (by exception up to 170 kV in some countries) are taken to be non-effectively earthed with the first-pole-to-clear factor $k_{pp} = 1.5$ pu, and the TRVs are all underdamped and represented by two-parameter TRVs. Above 100 kV, power systems are considered to be effectively earthed with $k_{pp} = 1.3$ pu (except for transformer-fed faults covered by test duty T10 where $k_{pp} = 1.5$ pu), and the TRVs are overdamped for test duties T100 and T60 (represented by four parameters) and underdamped for test duties T30 and T10.

The influence of added R, L and C components on standard TRVs is examined for a 245 kV (L and C) and a 1100 kV (R only) circuit breaker:

- The addition of capacitance (in the form of a surge capacitor or a shunt capacitor bank) will decrease the rate of rise of the recovery voltage and increase the peak value of the TRV driving the TRVs in the direction of underdamping if initially overdamped and more underdamped if already underdamped.
- The addition of resistance (in the form of opening resistors) has the effect of driving the TRVs in the direction toward overdamping. The effect tends to be greatest for the underdamped TRVs and minimal for the overdamped TRVs unless the resistor value is very low.
- The addition of inductance (in the form of a series reactor) adds to the TRVs in that the reactor and associated stray capacitance produces its own TRV. The latter TRV is high frequency, and mitigation is usually required by the addition of a surge capacitor.

Out-of-phase switching is treated as an extension of the terminal fault cases and represents the event in which a circuit breaker connecting two systems is inadvertently closed when the systems are not synchronized. The fault current level is standardized at 25% of the circuit breaker short-circuit rating and the TRV is a four-parameter one with the lowest rate of rise but the highest peak value of all terminal fault TRVs.

Asymmetrical currents will occur when, for example, a circuit breaker closes into a bolted three-phase fault. An asymmetrical current will occur on the phases where making is at a source voltage point-on-wave other than a voltage peak, the maximum asymmetry being for closing at a source voltage zero crossing. This case is less onerous for the circuit breaker from a TRV perspective (because of the shifted current zero crossing relative to the voltage) but is more onerous thermally because of the extended major loops of current.

Bibliography

1. IEC62271-100 (2012) *High-Voltage Switchgear and Controlgear—Part 100: Alternating Current Circuit-Breakers*, International Electrotechnical Commission.
2. IEC62271-306 (2012) *High-Voltage Switchgear and Controlgear—Part 306: Guide to IEC 62271-100, IEC 62271-1 and Other IEC Standards Related to Alternating Current Circuit-Breakers*, International Electrotechnical Commission.
3. IEEEC37.011 (2011) *IEEE Application Guide for Transient Recovery Voltage for AC High-Voltage Circuit Breakers*, Institute of Electrical and Electronics Engineers.
4. Johnson, K.S. (1927) *Transmission Circuits for Telephonic Communication*, D. Van Nostrand Company, Inc. (first published in 1924).
5. Boehne, E.W. (1935) The determination of circuit recovery rates. *AIEE Trans.*, **54**, 530–539.
6. Hammarlund, P. (1946) Transient recovery voltage subsequent to short-circuit interruption with special reference to Swedish power system. Proceedings No. 189, Royal Swedish Academy of Engineering Sciences.
7. Colclasser, R.G. and Buettner, D.E. (1969) The travelling-wave approach to transient recovery voltage. *IEEE Trans. Power Apparatus Syst.*, 1028–1035.
8. Robert, D., Martin, F. and Taisne, J.P. (2007) Insertion of current limiting or load sharing reactors in a substation: impact on the specifications. CIGRE 44, Bruges.
9. Greenwood, A. (1971) *Electrical Transients in Power Systems*, John Wiley & Sons, Inc.

5

Short-Line Faults

5.1 General

The short-line fault is so called because the fault in this case occurs at a short distance along a transmission line from the circuit breaker. The distance of interest is in the order of hundreds of metres to 1 km or so, and the fault is sometimes referred to as a kilometric fault.

The circuit for this case is shown in Figure 5.1, where V_S is the source voltage peak line to ground voltage, X_S is the source impedance, V_L is the peak voltage at the circuit breaker line-side terminal at current interruption and X_{eff} and Z_{eff} are the effective line and surge impedances, respectively. The fault current I_L is usually expressed as a percentage of the circuit breaker short-circuit current rating I_S. A 90% short-line fault thus means that $I_L = 0.9 I_S$.

The circuit breaker transient recovery voltage (TRV) is the difference between the source-side TRV and the line-side transient. At current interruption, the voltage on the line side is V_L with a decreasing linear voltage profile to zero along the line to the fault location as shown in Figure 5.2. The line-side voltage must go to zero, but it does not ring down in the way of an oscillatory circuit because now we are dealing with distributed rather than the lumped elements discussed earlier in Chapters 2 and 4. Instead, the line voltage goes to zero by means of successive travelling waves as shown in the next section.

5.2 Line-Side Voltage Calculation

The initial condition for the line-side voltage is as shown in Figure 5.2. The subsequent travelling wave must travel in both directions, and this is achieved by half of the wave going to the left and the other half going to the right (Figure 5.2). For ease of illustration, V_L is taken as 2 pu, and each wave is thus 1 pu peak at time zero. The progression of the right-going wave is shown in Figure 5.3a–c in time steps of one-quarter of the line travelling time. Likewise, the progression of the left-going wave is shown in Figure 5.4a–c in the same time steps. This wave is immediately totally reflected at the open circuit breaker and essentially follows the right-going wave with a reversed profile. The voltage profile on the line at any point in time is determined by summing the individual wave voltages. The voltage at the fault location must always sum to zero.

After twice the line travel time t_L, the two waves have returned to the circuit breaker with reversed polarity as shown in Figure 5.5. The process starts over again, and the two waves will return after the same time interval with positive polarity. If we take the voltage at the circuit

---

Current Interruption Transients Calculation, First Edition. David F. Peelo.
© 2014 John Wiley & Sons, Ltd. Published 2014 by John Wiley & Sons, Ltd.

Short-Line Faults

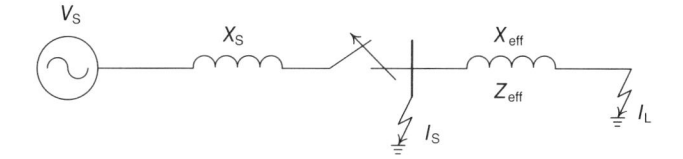

Figure 5.1 Short-line fault circuit representation.

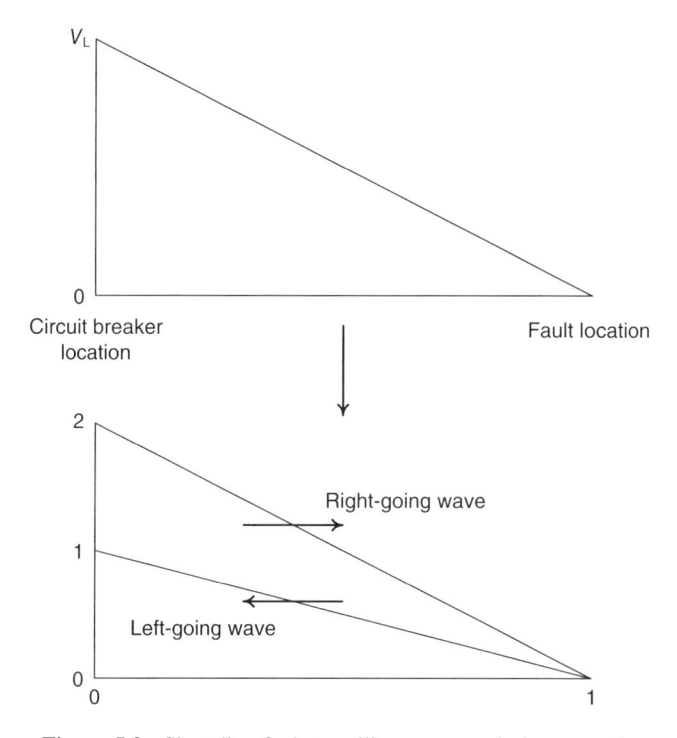

Figure 5.2 Short-line fault travelling wave analysis approach.

breaker at each instant in time and plot it against travel time, the voltage has the sawtooth-type waveshape shown in Figure 5.6. Viewed in oscillatory circuit terms, the starting point of the transient is V_L, the axis of oscillation is zero and the peak value is $2V_L$. This, of course, assumes no damping or other considerations, and the actual peak is given by kV_L, where k is a so-called peak factor, in effect another quasi-amplitude factor.

With reference to Figure 5.1, we can write

$$\begin{aligned} V_L &= V_S - X_S I_L \\ &= V_S - \frac{V_S}{I_S} I_L \\ &= V_S \left(1 - \frac{I_L}{I_S}\right) \\ &= V_S (1 - M), \end{aligned}$$

where M is I_L/I_S.

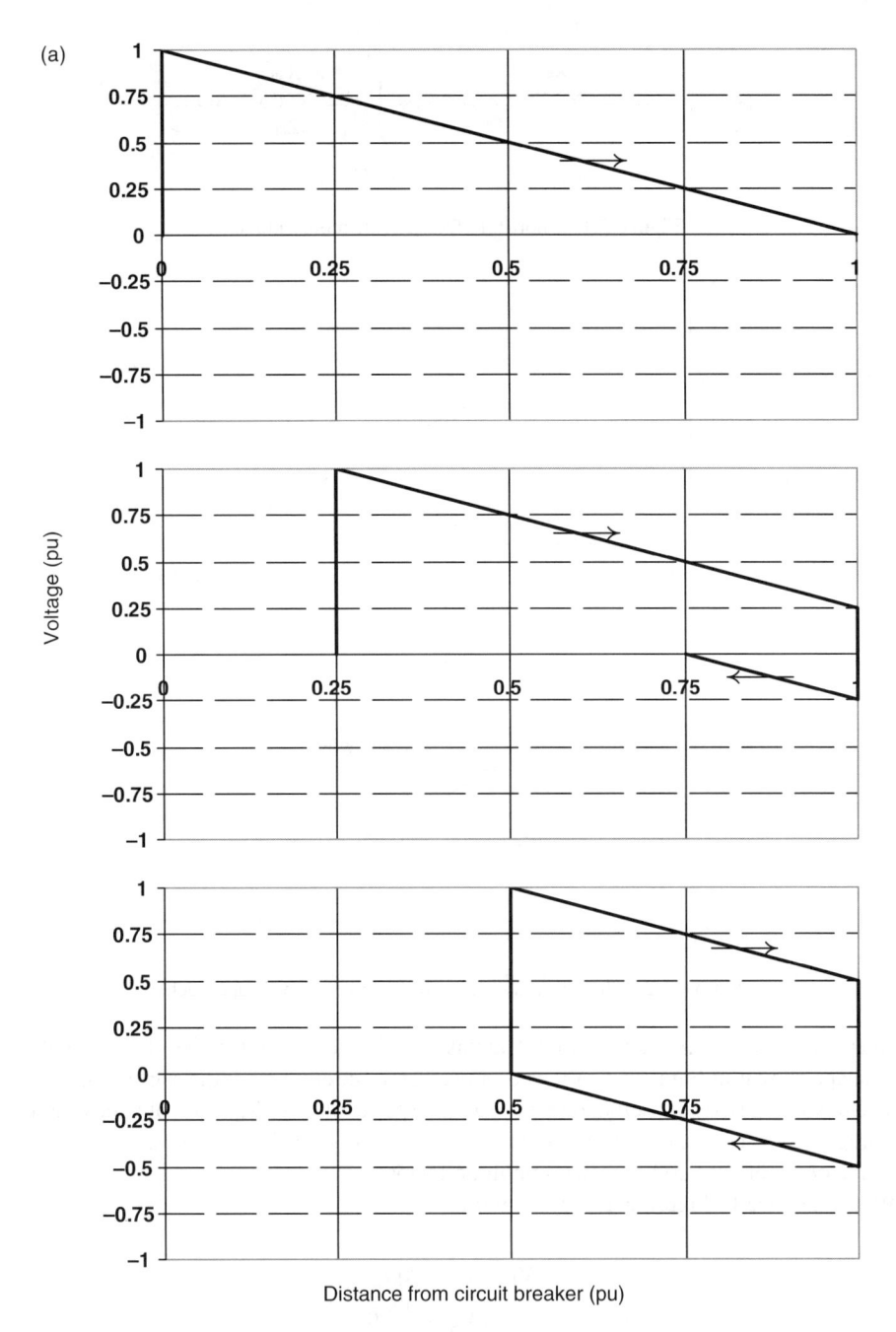

Figure 5.3 (a–c) Short-line fault right-going travelling wave.

(b)

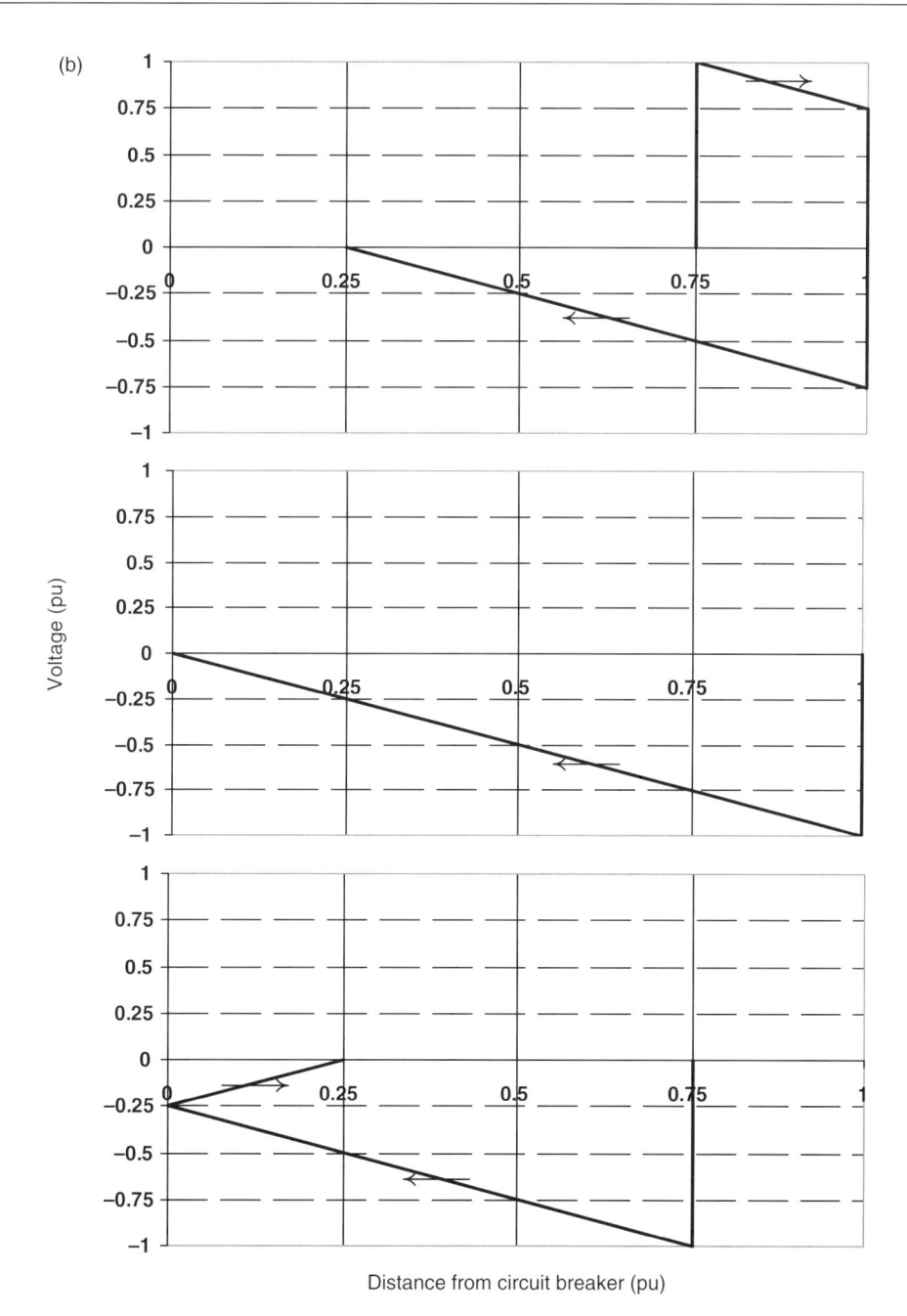

Voltage (pu)

Distance from circuit breaker (pu)

Figure 5.3 (*Continued*)

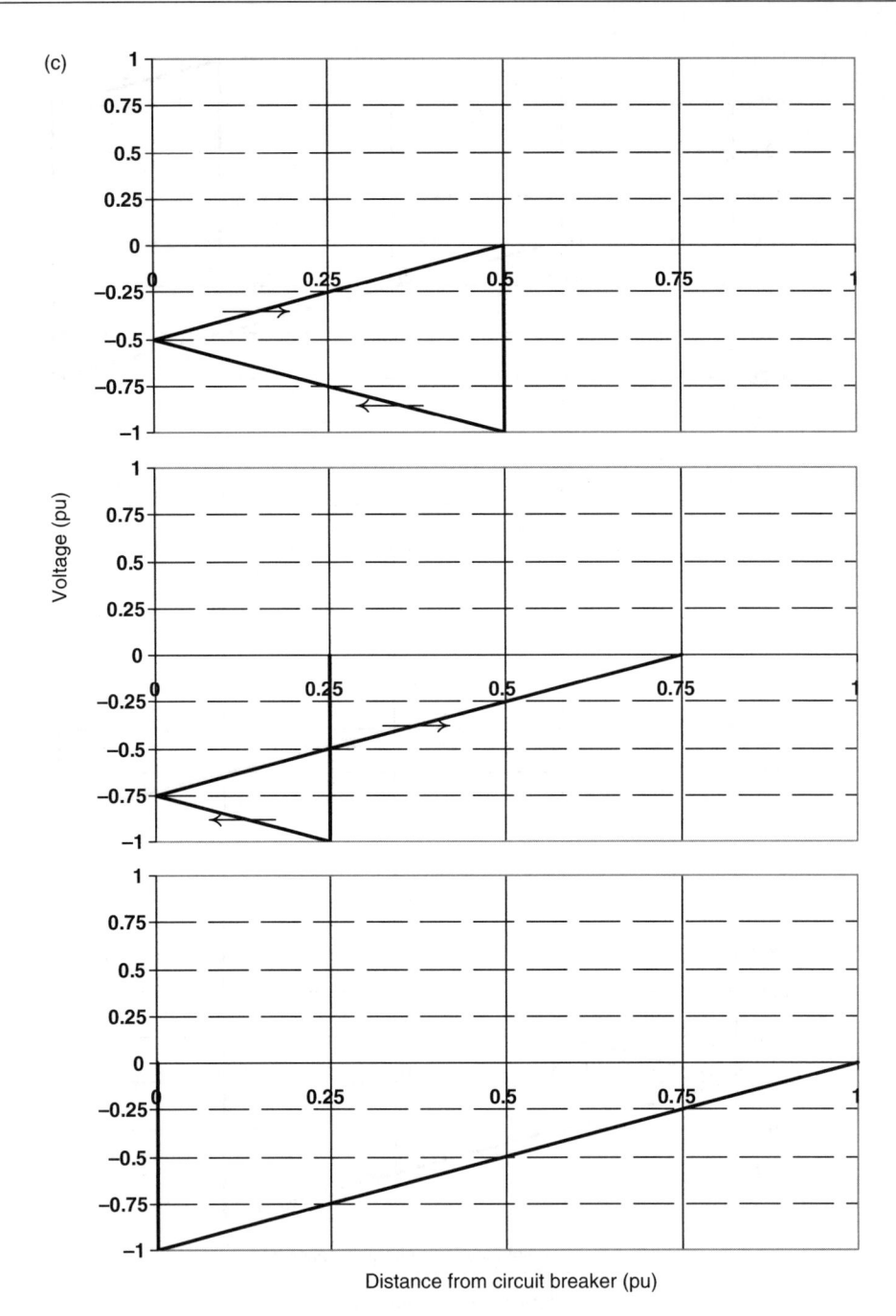

Figure 5.3 (*Continued*)

Short-Line Faults

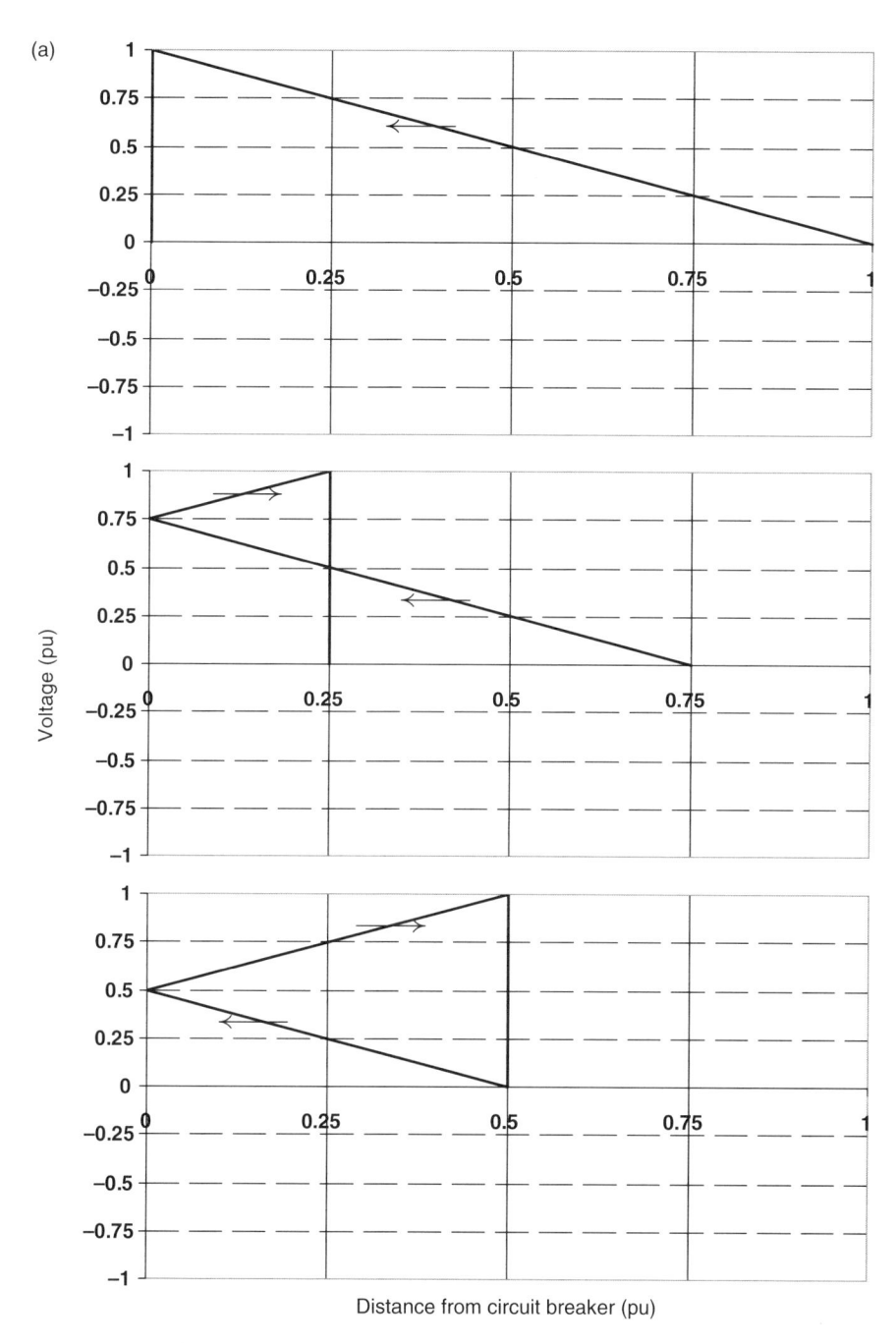

Figure 5.4 (a–c) Short-line fault left-going travelling wave.

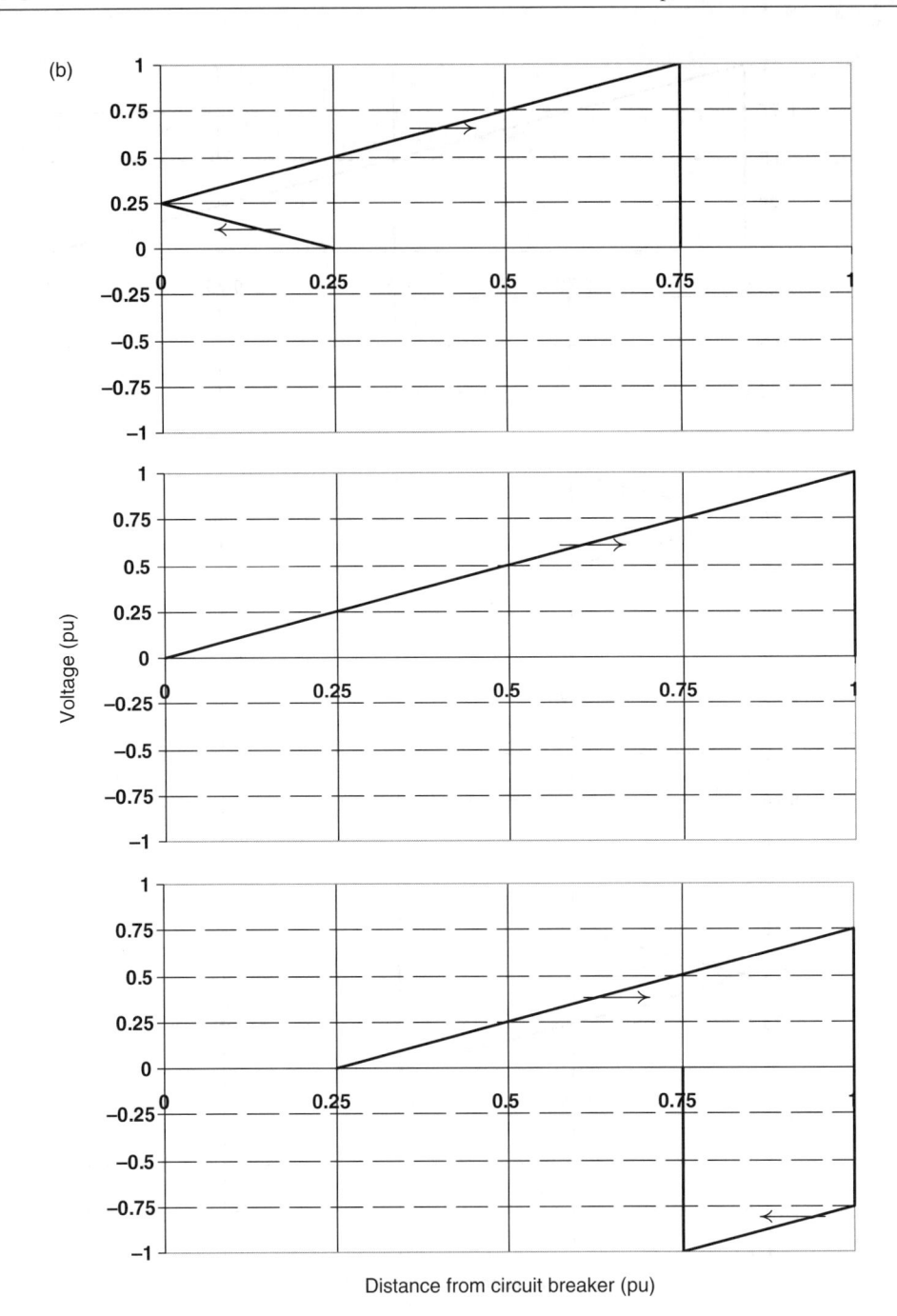

Figure 5.4 (*Continued*)

Short-Line Faults

(c)

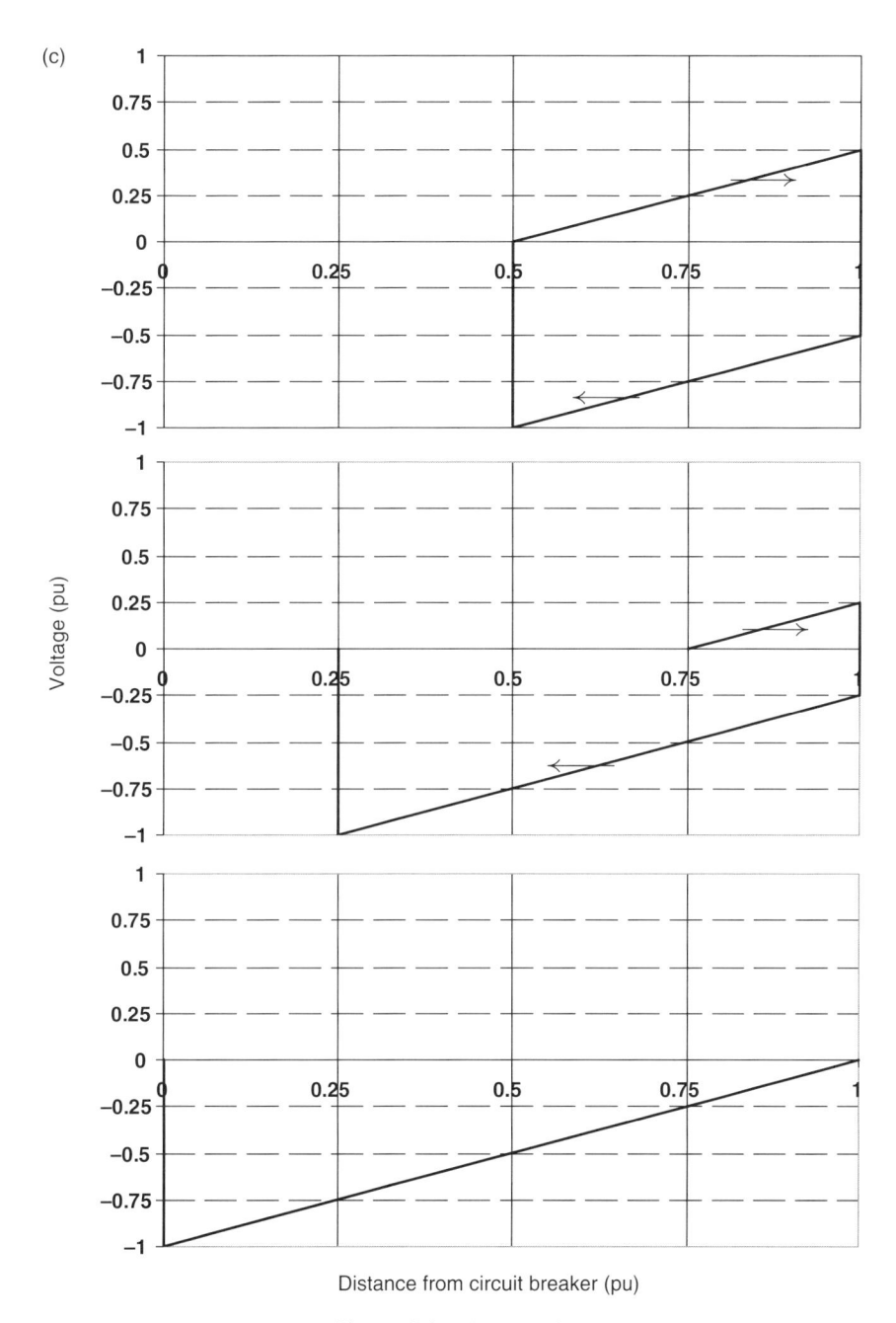

Figure 5.4 (*Continued*)

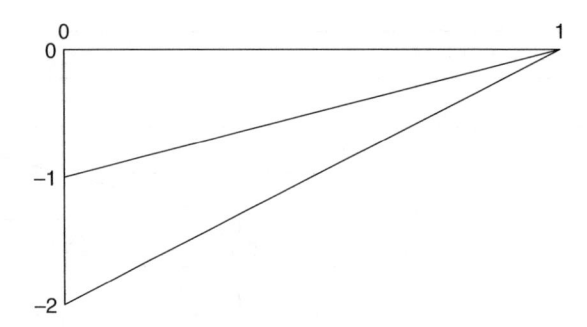

Figure 5.5 Voltage profile on the line after twice the line travelling time.

The peak value of the line-side transient V_{pk} is

$$V_{pk} = k(1-M)V_S. \tag{5.1}$$

The rate of rise of recovery voltage (RRRV) of the line-side transient is given by

$$\begin{aligned} \text{RRRV} &= \frac{V_{pk}}{2t_L} \\ &= Z_{eff}\frac{di}{dt} \\ &= \sqrt{2}\omega I_L Z_{eff}. \end{aligned} \tag{5.2}$$

To now calculate the value of k, readers can readily verify that the sequence network for a single-line-to-ground fault is as shown in Figure 5.7. The sequence impedances are in series and Z_{eff} is given by

$$\begin{aligned} Z_{eff} &= \frac{Z_1 + Z_2 + Z_0}{3} \\ &= \frac{2Z_1 + Z_0}{3} \quad \text{for } Z_1 = Z_2. \end{aligned}$$

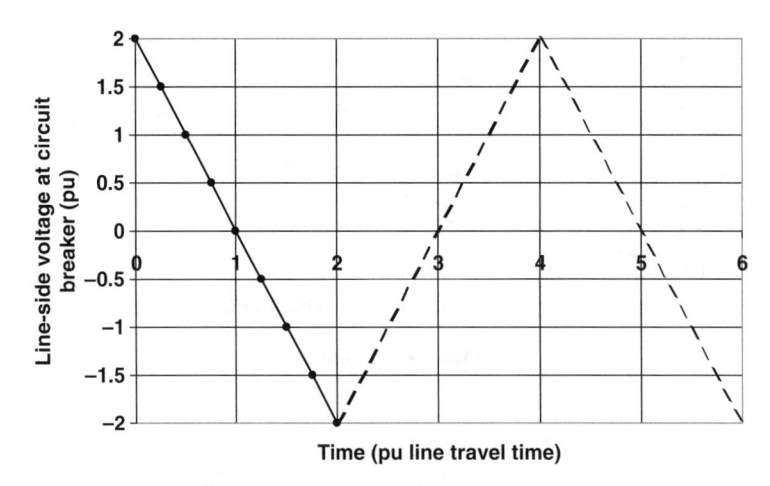

Figure 5.6 Voltage at circuit breaker as a function of travel time.

Short-Line Faults

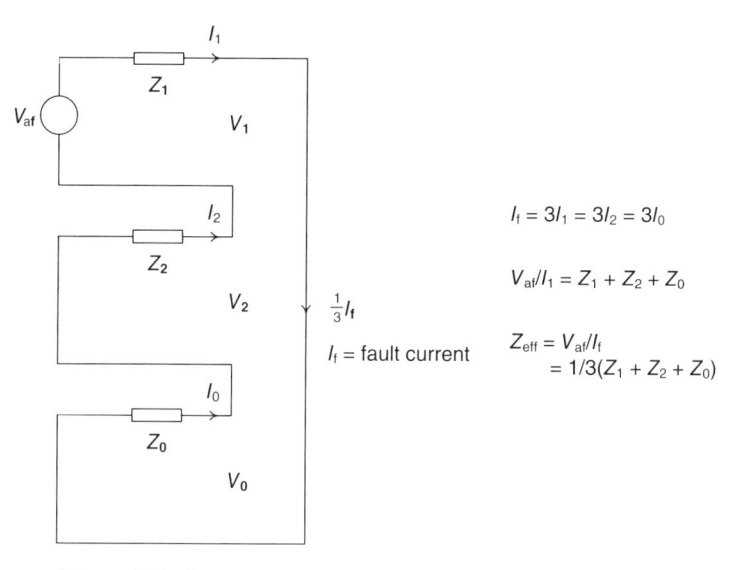

Figure 5.7 Sequence network for a single-line-to-ground fault.

$I_f = 3I_1 = 3I_2 = 3I_0$

$V_{af}/I_1 = Z_1 + Z_2 + Z_0$

I_f = fault current

$Z_{eff} = V_{af}/I_f$
$= 1/3(Z_1 + Z_2 + Z_0)$

Likewise, ignoring resistance and considering only reactance and thereby inductance both per unit length of line, we can write

$$L_{eff} = \frac{2L_1 + L_0}{3}.$$

From Eq. (5.2),

$$V_{pk} = 2t_L\sqrt{2}\omega I_L \left(\frac{2Z_1 + Z_0}{3}\right),$$

and V_L is given by

$$V_L = \sqrt{2}I_L\omega\left(\frac{2L_1 + L_0}{3}\right)l,$$

where l is the distance to the fault. Taking v as the velocity of the travelling wave, k is calculated as follows:

$$\begin{aligned}
k &= \frac{V_{pk}}{V_L} \\
&= 2\frac{t_L}{l}\left(\frac{2Z_1 + Z_0}{2L_1 + L_0}\right) \\
&= \frac{2}{v}\left(\frac{2Z_1 + Z_0}{2L_1 + L_0}\right) \\
&= \frac{2Z_1(2 + Z_0/Z_1)}{v L_1(2 + L_0/L_1)}.
\end{aligned} \quad (5.3)$$

Now we will make the assumption that the high frequency and power frequency per unit length inductances (L_1) and capacitances (C_1) of the line are approximately equal.

$$Z_1 = \sqrt{\frac{L_1}{C_1}}$$

and

$$v = \frac{1}{\sqrt{L_1 C_1}},$$

giving $Z_1/L_1 = v$ and Eq. (5.3) becomes

$$k = \frac{2(2 + Z_0/Z_1)}{2 + L_0/L_1}.$$

Taking $L_0/L_1 = 3$ and $Z_0/Z_1 = 2$, both conservative values, $d = 1.6$, and Eq. (5.1) becomes

$$V_{pk} = 1.6(1 - M)V_S. \tag{5.4}$$

To illustrate the use of Eqs. (5.2) and (5.4), we will take the case of a 245 kV, 40 kA circuit breaker at 50 Hz and consider 90, 75 and 60% faults. In the IEC circuit breaker standard, these test duties are designed as L90, L75 and L60, respectively. Also, the standard specifies that $Z_{eff} = 450\,\Omega$.

For L90, we can write

$$\text{RRRV} = \sqrt{2} \times 314 \times 36 \times 450 \times 10^{-6}\,\text{kV}/\mu\text{s}$$
$$= 7.2\,\text{kV}/\mu\text{s},$$

$$V_{pk} = 1.6 \times 0.1 \times \frac{245\sqrt{2}}{\sqrt{3}}\,\text{kV}$$
$$= 32\,\text{kV}.$$

$$\text{Time to peak, } t_p = \frac{32}{7.2}$$
$$= 4.44\,\mu\text{s}.$$

$$\text{Distance to the fault, } l = 300 \cdot \frac{t_p}{2}$$
$$= 666\,\text{m},$$

where 300 m/μs is the speed of light.

The values for all the above cases are given in Table 5.1 and plotted in Figure 5.8.

Short-Line Faults

Table 5.1 L90, L75 and L60 line-side transient characteristics for a 245 kV, 40 kA, 50 Hz circuit breaker.

Test duty	V_{pk} (kV)	RRRV (kV/μs)	Time to peak (μs)	Distance to fault (m)
L90	32	7.2	4.44	666
L75	80	6	13.33	2000
L60	128	4.8	26.67	4000

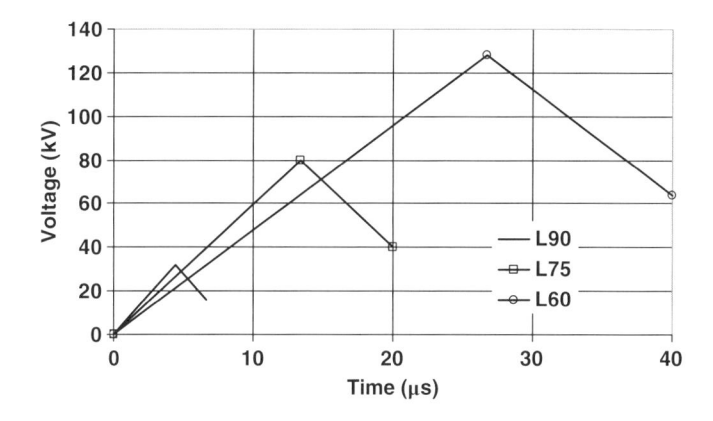

Figure 5.8 L90, L75 and L60 line-side transients for a 245 kV, 40 kA, 50 Hz circuit breaker.

Table 5.2 L90 line-side transient characteristics for 245 kV, 50 Hz circuit breakers rated at 31.5, 40 and 50 kA.

Short-circuit current rating (kA)	V_{pk} (kV)	RRRV (kV/μs)	Time to peak (μs)	Distance to fault (m)
31.5	32	5.66	5.65	847
40	32	7.2	4.44	666
50	32	9	3.55	533

A second case to consider is for a 245 kV circuit breaker at 50 Hz and varying short-circuit current ratings of 31.5, 40 and 50 kA. The case results are given in Table 5.2 and plotted in Figure 5.9.

5.3 Effect of Added Capacitance

The effect of adding extra capacitance line to earth on the line side of the circuit breaker is to reduce the rate of rise of the line-side transient. The peak value of the transient is unaffected because it relates only to the source voltage and the fault current (refer to Eq. (5.4)).

The effect of added capacitance C_{add} can be estimated by treating the line-side transient as an undamped oscillation based on the line inductance L and an equivalent capacitance C_{eq}. With reference to Figure 5.6 and ignoring damping, that is, $V_{pk} = 2V_L$, the time to peak t_p is given by

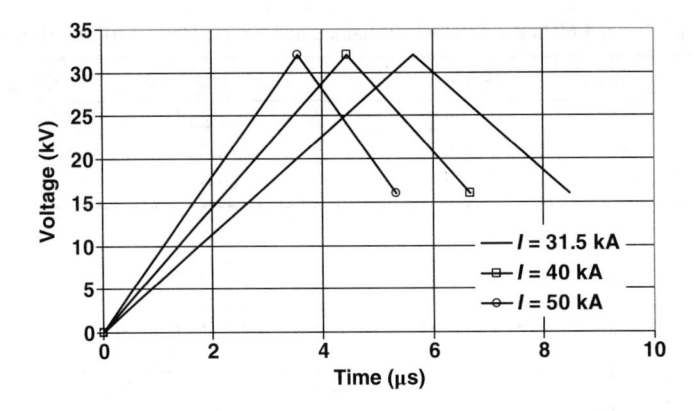

Figure 5.9 L90 line-side transients for a 245 kV, 50 Hz circuit breakers rated at 31.5, 40 and 50 kA.

$$t_p = \pi\sqrt{LC_{eq}}. \tag{5.5}$$

t_p is also given by

$$\begin{aligned} t_p &= \frac{V_{pk}}{\text{RRRV}} \\ &= \frac{2\sqrt{2}\omega IL}{\sqrt{2}\omega IZ_{eff}} \\ &= 2L\sqrt{\frac{C_L}{L}} \\ &= 2\sqrt{LC_L}, \end{aligned} \tag{5.6}$$

where C_L is the total line capacitance.

Equating Eqs. (5.5) and (5.6) and solving for C_{eq},

$$\begin{aligned} C_{eq} &= \frac{4}{\pi^2}C_L \\ &= 0.4 C_L. \end{aligned}$$

The time to peak for the base no added capacitance case is

$$t_{pb} = \pi\sqrt{0.4 LC_L}$$

and for the added capacitance case

$$t_{pa} = \pi\sqrt{L(0.4C_L + C_{add})}.$$

Short-Line Faults

The reduction in the rate of rise of the line-side transient due to added capacitance is the ratio of

$$t_{pb}/t_{pa} = \sqrt{\frac{0.4C_L}{0.4C_L + C_{add}}}$$

$$= \sqrt{\frac{C_L}{C_L + 2.5C_{add}}}.$$

Equation (5.2) can now be used to derive the equation for the rate of rise with added capacitance, giving

$$\text{RRRV} = \sqrt{2}\omega I Z_{eff}\sqrt{\frac{C_L}{C_L + 2.5C_{add}}}. \tag{5.7}$$

Taking the same 245 kV, 40 kA, 50 Hz circuit breaker, we will calculate the effect of added capacitance for the L90 case. The base case with no added capacitance thus has $V_{pk} = 32$ kV and RRRV = 7.2 kV/μs. To calculate the value for C_L, the line impedance X_L is given by

$$X_L = \frac{V_{pk}}{1.6\sqrt{2}I}$$
$$= 0.393\,\Omega.$$

Alternatively,

$$X_L = \frac{245}{36\sqrt{3}} - \frac{245}{40\sqrt{3}}$$
$$= 3.929 - 3.536$$
$$= 0.393\,\Omega,$$
$$L = 1.25\,\text{mH},$$
$$C_L = \frac{1.25 \times 10^{-3}}{450^2}$$
$$= 6.17\,\text{nF}.$$

For $C_{add} = 1$ nF,

$$\text{RRRV} = 7.2\sqrt{\frac{6.17}{6.17 + 2.5}}$$
$$= 6\,\text{kV/μs}.$$

In Table 5.3, the RRRV values are given for added capacitances of 1, 2, 5 and 10 nF and plotted in Figure 5.10.

Applying dedicated capacitor units for the purpose of influencing the line-side transient is not normal utility practice. However, there have been incidents when such capacitors were added after the failure of a circuit breaker to clear a short-line fault. For maximum effect, the capacitors have to be placed as close as possible to the circuit breaker.

Table 5.3 L90 line-side transient characteristics for a 245 kV, 40 kA, 50 Hz circuit breaker with varying added line to earth capacitance.

C_{add} (nF)	V_{pk} (kV)	RRRV (kV/μs)	Time to peak (μs)
0	32	7.2	4.44
1	32	6.07	5.27
2	32	5.35	5.98
5	32	4.14	7.73
10	32	3.2	9.98

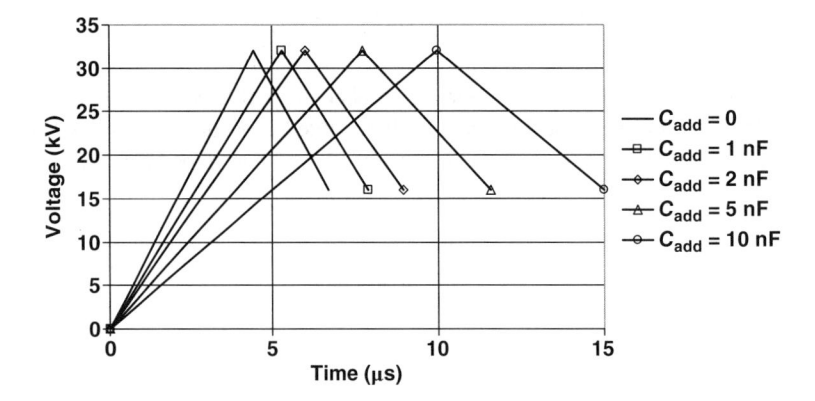

Figure 5.10 L90 line-side transients for a 245 kV, 40 kA, 50 Hz circuit breaker with varying added line to earth capacitance.

5.4 Discussion

The short-line fault case is characterized by a fast-rising line-side transient. The case is particularly onerous for circuit breakers subject to thermal re-ignitions.

The short-line case is treated as a single-phase-to-earth fault event in the standards. The reason for this is that the RRRV is higher for the third-pole-to-clear than for the first-pole-to-clear. From Section 3.5 and Figure 3.20, the circuit representation for the first and third pole clearing is shown in Figure 5.11. Using current injection, the effective surge impedance for the first-pole-to-clear Z_{eff1} is given by

$$Z_{eff1} = \frac{3Z_0 Z_1}{Z_1 + 2Z_0}$$

and for the third-pole-to-clear Z_{eff3} by

$$Z_{eff3} = \frac{2Z_1 + Z_0}{3}.$$

In general, $Z_0 > Z_1$, and the ratio of Z_0/Z_1 is such that $Z_{eff3} > Z_{eff1}$.

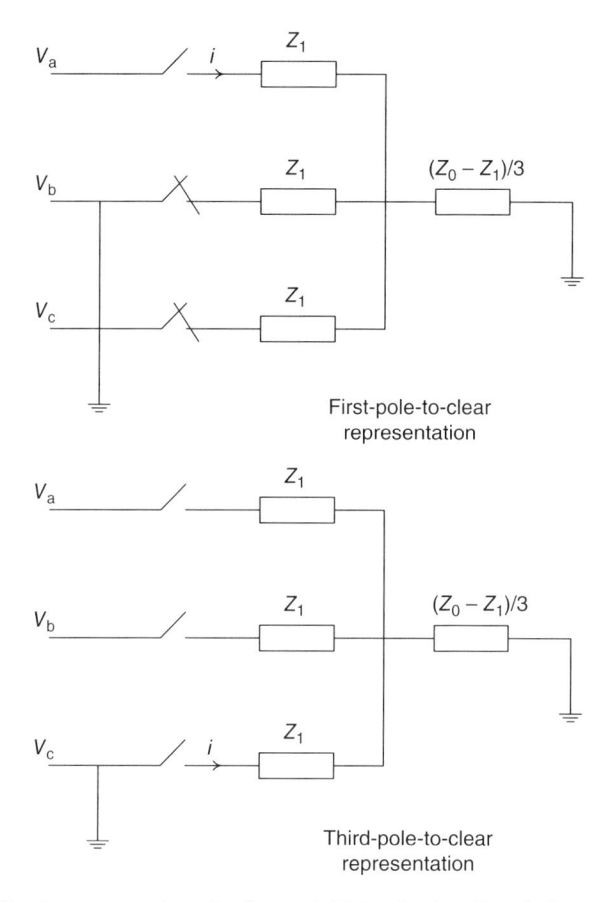

Figure 5.11 Circuit representations for first and third pole short-line fault current interruption.

The test fault levels are in the T100 to T60 range, and the source-side TRVs are those for the terminal faults but with $k_{pp} = 1$. Note also that the rise time of the line-side transients falls within the time delay period of the source-side TRV and its contribution to the total voltage across the circuit breaker is minimal to insignificant.

Bibliography

1. Thoren, B. (1966) Short line faults. *Elteknik*, **9** (2), 25–33.
2. van der Sluis, L. (2001) *Transients in Power Systems*, John Wiley & Sons, org.
3. Janssen, A.L.J., Kapetanovic, M., Peelo, D.F., Smeets, R.P.P. and van der Sluis, L. (2014) *Switching in Electric Transmission and Distribution Systems*, John Wiley & Sons, org.
4. IEC62271-306 (2012) *High-Voltage Switchgear and Controlgear – Part 306: Guide to IEC 62271-100, IEC 62271-1 and Other IEC Standards Related to Alternating Current Circuit-Breakers*, International Electrotechnical Commission.
5. IEEEC37.011 (2011) *IEEE Application Guide for Transient Recovery Voltage for AC High-Voltage Circuit Breakers*, Institute of Electrical and Electronics Engineers.

6

Inductive Load Switching

6.1 General

The inductive load switching cases to be considered are the switching of shunt reactors and unloaded transformers. Shunt reactor loads are 90° lagging in a number of load circuit configurations and represent an onerous switching duty for circuit breakers. Unloaded transformer switching also presents a lagging power factor but less than 90° and is generally not a significant test duty for circuit breakers. However, particularly in North America, air-break disconnectors (also known as disconnect switches) are used to switch out unloaded transformers, and this represents a special application case.

For the vast majority of shunt reactor installations, the circuit breaker is applied on the source side of the shunt reactor, and variations with respect to the reactor neutral earthing are as shown in Figure 6.1. The variation in the shunt reactor neutral earthing is related to system voltages but also to individual utility requirements:

- Shunt reactors with isolated neutrals are most common at system voltages below 72.5 kV but are also in use up to 170 kV in some countries.
- Shunt reactors with earthed neutrals are in use at all system voltage levels.
- Shunt reactors with neutrals earthed through a neutral reactor are mostly in use at voltages of 300 kV and above in order to enable single-phase fault clearing and reclosing.

Each of the above variations represents a unique case from a transient analysis perspective. The variations can, however, be treated by considering a general case and then adapting the general solution to the case in question. A further and separate consideration is when the circuit breaker is on the neutral side of the shunt reactor as may be the case at system voltages below 72.5 kV.

The three-phase circuit for the general case is shown in Figure 6.2. Setting L_n to infinity or zero will give the isolated and earthed neutral cases, respectively. As for the fault cases studied earlier, the first-pole-to-clear will experience the highest transient recovery voltage (TRV), and we need to derive the corresponding circuit.

Current Interruption Transients Calculation, First Edition. David F. Peelo.
© 2014 John Wiley & Sons, Ltd. Published 2014 by John Wiley & Sons, Ltd.

Inductive Load Switching

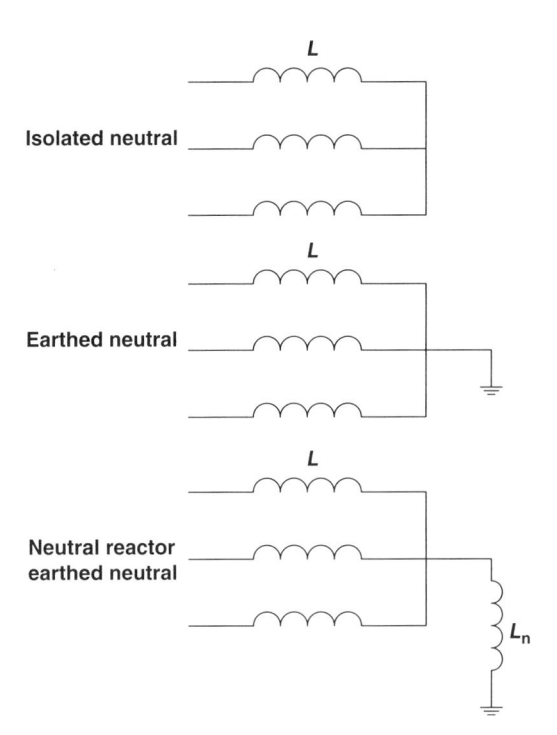

Figure 6.1 Shunt reactor load circuit variations.

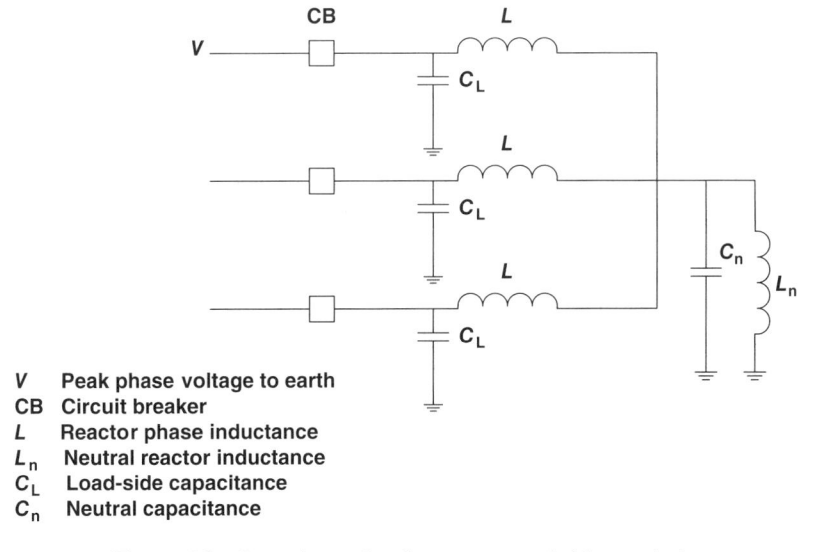

V   Peak phase voltage to earth
CB Circuit breaker
L   Reactor phase inductance
L_n Neutral reactor inductance
C_L Load-side capacitance
C_n Neutral capacitance

Figure 6.2 General case for shunt reactor switching analysis.

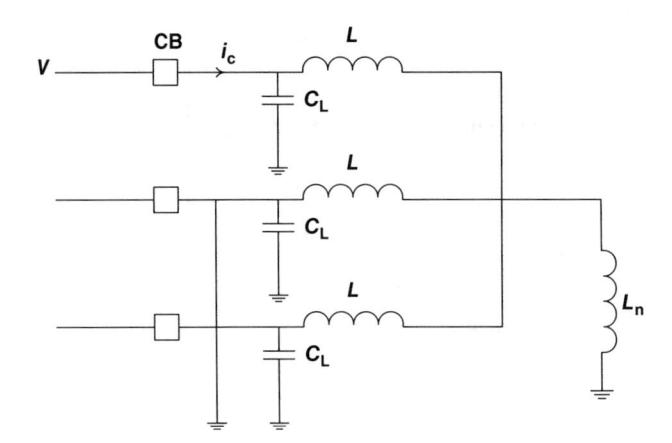

Figure 6.3 First-pole-to-clear considerations.

Applying the principle of superposition only, the current in the first-pole-to-clear needs to be considered, and we can conclude the following:

- The current path is first through the first phase inductance and then splits between the other two phase inductances and the neutral inductance in parallel. The current in the former two phases can be taken as flowing into an infinite bus at earth potential as shown in Appendix F.
- Before current interruption, the circuit is balanced, and the neutral point is at zero potential. No energy is therefore stored in C_n, and it can be ignored in the transient calculation.

Figure 6.3 shows the circuit taking the above considerations into account. The circuit is further simplified in Figure 6.4 first as shown in the upper version and then as shown in the lower final version. After the current in the first pole is interrupted, the circuit will become unbalanced, and a voltage will appear on the neutral point in the same way as for the first-pole-to-clear in a three-phase ungrounded fault. The neutral voltage is given by the factor K, which equals 0.5 pu for the isolated neutral case and zero for the earthed neutral case. Given that the load inductance is $L(1+K)$, V_{fptc} has to be $V(1+K)$ in principle and to maintain the correct current magnitude, V being the peak value of the phase voltage to earth.

Shunt reactor switching has unique features compared with other current interrupting duties The features in question are current chopping and the significance of arc voltage:

- Current chopping is essentially interruption of the current before a natural zero crossing. The chopping is caused by an unstable interaction between the arc and the capacitance in parallel with the circuit breaker as the current approaches zero. A negatively damped oscillation is established, which increases in magnitude with time and eventually crosses the zero line, thus interrupting the current. The net effect of the current chopping is that it leaves a trapped current and therefore trapped energy in the shunt reactor, which can contribute to subsequent transient overvoltages.
- In contrast to fault current interruption in which the arc voltage is not significant compared with the system voltage, this voltage is a consideration in shunt reactor switching. At the

Inductive Load Switching

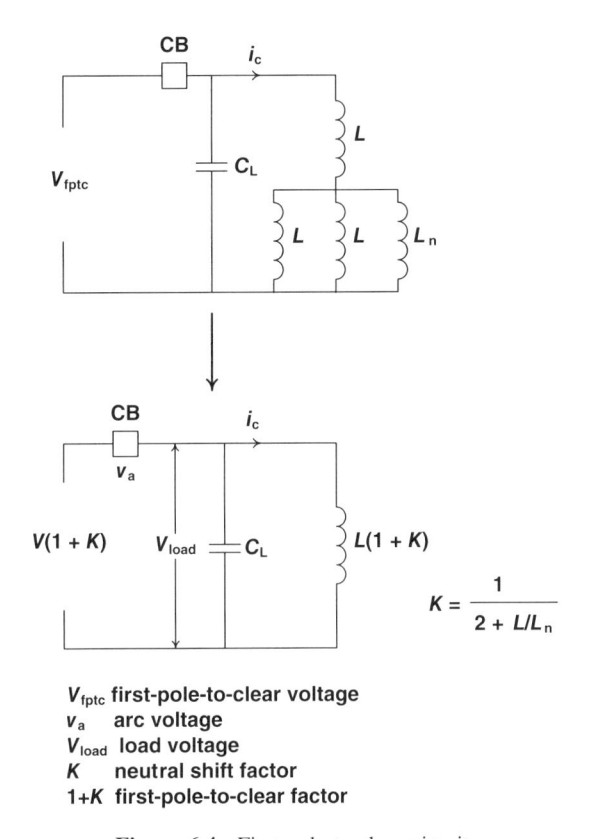

Figure 6.4 First-pole-to-clear circuit.

V_{fptc} first-pole-to-clear voltage
V_a arc voltage
V_{load} load voltage
K neutral shift factor
$1+K$ first-pole-to-clear factor

$$K = \frac{1}{2 + L/L_n}$$

relevant lower current values, the arc voltage is high enough to contribute to the transient overvoltage magnitude. The arc voltage is in phase with the current and therefore of opposite polarity to the system voltage (refer to Figure 3.21). Seen from the perspective of the load, the "source" voltage at current interruption will equal $V(1+K) + v_a$ (refer to Figure 6.4).

The discussion to this point of shunt reactor switching is illustrated in Figure 6.5. The upper trace shows the current through the circuit breaker, and the lower trace shows the load-side voltage. The initial peak is known as the suppression peak overvoltage, and the subsequent load-side oscillation is an underdamped series oscillation as discussed in Chapter 2. The circuit breaker TRV is the difference between the source-side power frequency voltage and the load-side voltage.

6.2 General Shunt Reactor Switching Case

The quantity that we first need to calculate is the value of the suppression peak overvoltage. This is achieved by considering energy stored in the circuit at the instant of current chopping, that is, the energy stored in C_L and $L(1+K)$ (Figure 6.4). The maximum overvoltage will occur when all the available stored energy is stored in terms of voltage in C_L.

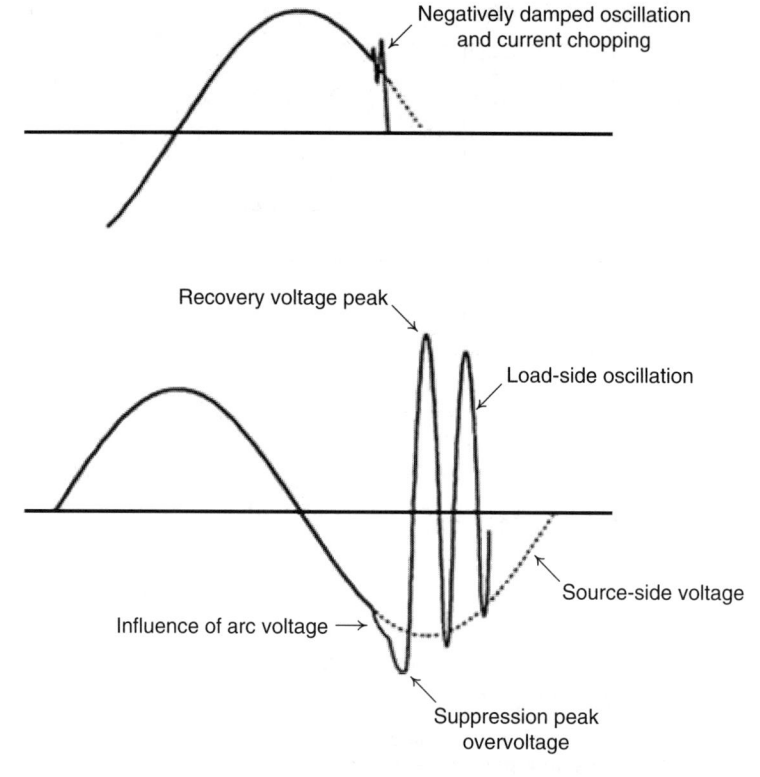

Figure 6.5 Shunt reactor switching phenomena.

At the instant just before current interruption, the so-called initial voltage V_{in} is given by

$$V_{in} = V + v_a. \qquad (6.1)$$

Expressing Eq. (6.1) in pu,

$$\frac{V_{in}}{V} = 1 + \frac{v_a}{V}$$

and defining $V_{in}/V = k_{in}$ pu

$$v_a = V(k_{in} - 1). \qquad (6.2)$$

From Figure 6.4,

$$\begin{aligned} V_{load} &= V(1+K) + v_a \\ &= V(1+K) + V(k_{in} - 1) \\ &= V(k_{in} + K). \end{aligned}$$

… Inductive Load Switching

The energy stored in C_L at the instant of current chopping is E_C:

$$E_C = 0.5 C_L [V(k_{in}+K)]^2. \tag{6.3}$$

Also from Figure 6.4, the chopped current level is i_c, and the energy stored in the inductance $L(1+K)$ is E_L:

$$E_L = 0.5 L(1+K) i_c^2. \tag{6.4}$$

From Eqs. (6.3) and (6.4) and defining the maximum overvoltage peak value as V_m, we can write

$$0.5 C_L V_m^2 = 0.5 C_L [V(k_{in}+K)]^2 + 0.5 L(1+K) i_c^2. \tag{6.5}$$

Readers can easily show that Eq. (6.5) simplifies to the per-unit form

$$k_b = \sqrt{(k_{in}+K)^2 + \frac{L(1+K)i_c^2}{C_L V^2}}, \tag{6.6}$$

where $k_b = V_m/V$.

Equation (6.6) is the suppression peak overvoltage relative to the shifted neutral, and the axis (aiming point) of the load-side oscillation is the shifted neutral. To calculate the suppression overvoltage peak value with respect to earth k_a in pu, we need to subtract K:

$$k_a = \sqrt{(k_{in}+K)^2 + \frac{L(1+K)i_c^2}{C_L V^2}} - K. \tag{6.7}$$

Equation (6.7) is the fundamental equation on which all further calculations of shunt reactor switching transients can be based.

The quantity that we need to calculate is the TRV across the circuit breaker. The load-side circuit, so far only considering $L(1+K)$ and C_L, also has losses represented by resistance R. We now have the series RLC case of Section 2.3 and generic oscillations as shown in Figure 2.4. We can now assign values to the vertical axis and to the time scale as shown in Figure 6.6. The oscillations are always underdamped as described by Eq. (2.22) because R has a low value to minimize losses.

To use the generic curves, the relationship of R/R_C to k_{af} needs to be known (refer to Figure 2.9). Equation (2.22) is the sum of two sinusoidal quantities having a maximum at π radians; therefore, we can write

$$\sqrt{1-d_s^2}\, t_g = \pi$$

or

$$t_g = \frac{\pi}{\sqrt{1-d_s^2}}.$$

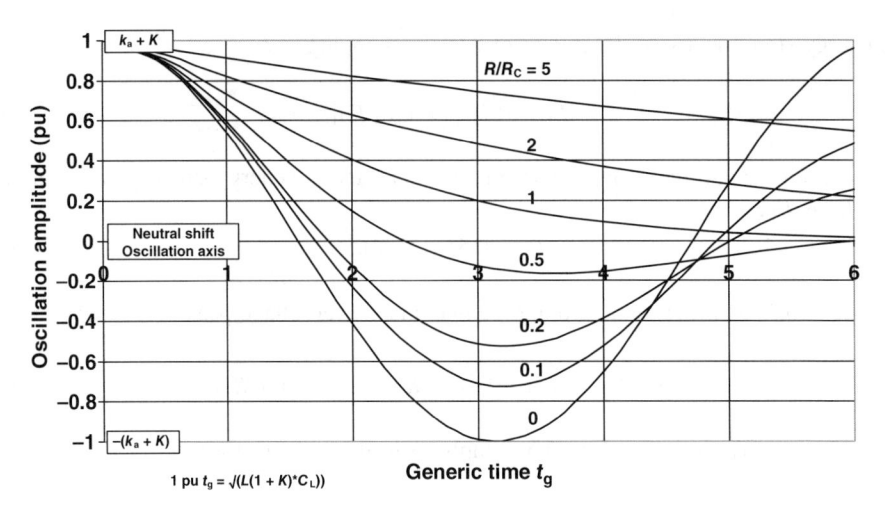

Figure 6.6 Load-side circuit generic oscillations.

Substituting in Eq. (2.22) to calculate $v_c(t)$ pu max,

$$v_c(t) \text{ pu max} = e^{-d_s\pi/\sqrt{1-d_s^2}}\left(\cos\pi + \frac{d_s}{\sqrt{1-d_s^2}}\sin\pi\right)$$

$$= e^{-d_s\pi/\sqrt{1-d_s^2}}(-1+0)$$

$$= -e^{-d_s\pi/\sqrt{1-d_s^2}}.$$

This is the peak value relative to the zero axis, and k_{af} is given by

$$k_{af} = 1 - \left(-e^{-d_s\pi/\sqrt{1-d_s^2}}\right)$$

$$= 1 + e^{-d_s\pi/\sqrt{1-d_s^2}}, \qquad (6.8)$$

where $d_s = R/R_C$.

The reverse equation, where k_{af} is known, is

$$d_s = \frac{1}{\sqrt{1 + (\pi/\ln(1/(k_{af}-1)))^2}}. \qquad (6.9)$$

Equations (6.8) and (6.9) have the format as Eqs. (2.44) and (2.46), respectively, which suggests the existence of a duality. This is discussed further in Section 7.2.1.2.

Shunt reactor circuits generally have low losses, and k_{af} is taken as 1.9 pu, the corresponding value of $d_s = R/R_C$ being 0.033 (Eq. (6.9)). On this basis, Eq. (2.22) becomes

$$v(t)_{pu} = e^{-0.033 t_g}(\cos t_g + 0.033 \sin t_g). \qquad (6.10)$$

To calculate the TRV across the circuit breaker, we will examine the relationship between the involved quantities as shown in Figure 6.7 in absolute values and in Figure 6.8 in per-unit

Inductive Load Switching

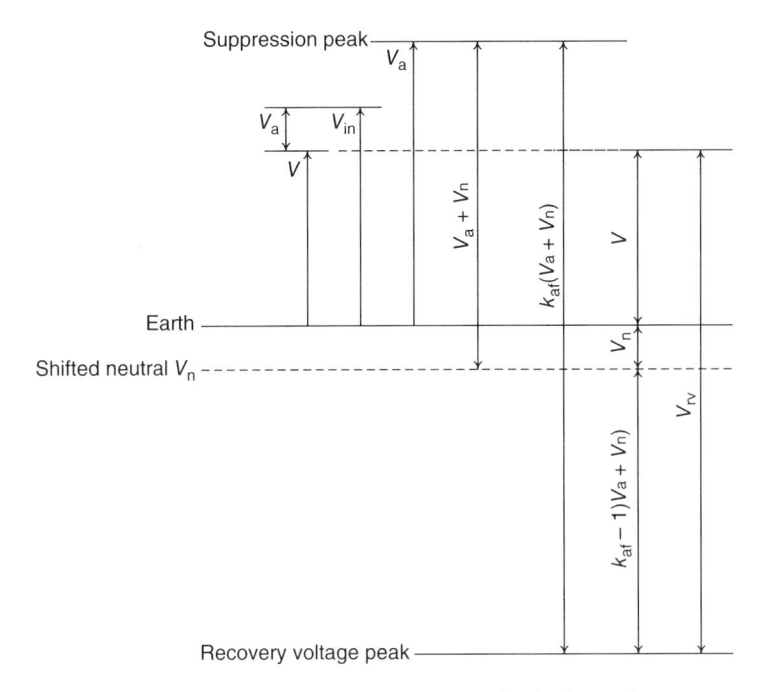

Figure 6.7 Current interruption voltages in absolute values.

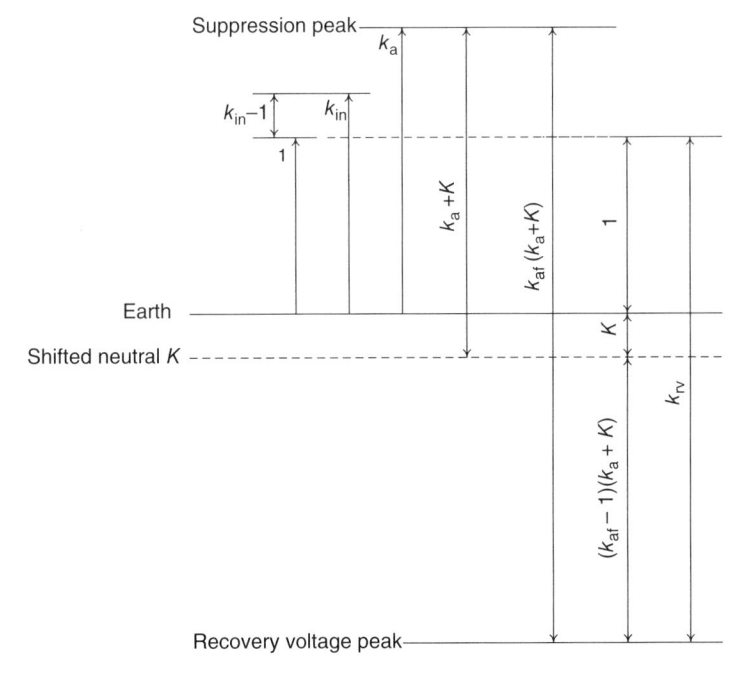

Figure 6.8 Current interruption voltages in per-unit values.

Table 6.1 Absolute and per-unit values.

Quantity	Values	
	Absolute (V)	Per unit
Source voltage peak	V	1
Initial voltage	V_{in}	k_{in}
Arc voltage	V_a	$k_{in} - 1$
Neutral shift voltage	V_n	K
TRV peak	V_{rv}	k_{rv}
Amplitude factor	—	k_{af}

values. Assume that the current is interrupted at a positive-going current zero crossing and thus at a positive source voltage peak. The absolute and per-unit values and their relationship are summarized in Table 6.1.

Because the frequency of the load-side transient is much higher than the power frequency, the TRV peak value V_{rv} can be calculated based on the peak value of the source voltage. The TRV is the difference between the source- and load-side voltages, and from Figure 6.7 we can write

$$V_{rv} = V + V_n + (k_{af} - 1)(V_a + V_n),$$

and rearranging

$$V_{rv} = V + k_{af} V_n + (k_{af} - 1) V_a. \tag{6.11}$$

Equation (6.11) is converted to the more useful per-unit version by dividing across by V:

$$\frac{V_{rv}}{V} = \frac{V}{V} + k_{af} \frac{V_n}{V} + (k_{af} - 1) \frac{V_a}{V},$$

giving

$$k_{rv} = 1 + k_{af} K + (k_{af} - 1) k_a. \tag{6.12}$$

As noted earlier, k_{af} is typically 1.9 pu, and a conservative approach that ignores damping is often used, that is, $k_{af} = 2$, and Eq. (6.12) becomes

$$k_{rv} = 1 + 2K + k_a. \tag{6.13}$$

Equation (6.7) is difficult to use because exact values of i_c may not be known; another approach is the method based on chopping numbers.[1] Rizk showed that the chopped current

---

[1] The chopping number approach is only applicable to oil, air-blast and SF_6 gas circuit breakers. It is not applicable to vacuum circuit breakers in which i_c is mainly dependent on contact material.

Table 6.2 Circuit breaker chopping numbers.

Circuit breaker type	Chopping number, λ (A/F$^{0.5}$)
Minimum oil	5.8×10^4 to 10×10^4
Air-blast	15×10^4 to 20×10^4
SF$_6$ gas	0.4×10^4 to 19×10^4

level is related to the capacitance in parallel with the circuit breaker by the equation

$$i_c = \lambda \sqrt{C_p}, \tag{6.14}$$

where i_c is the chopped current level in A, C_p is the total capacitance in parallel with the circuit breaker in F and λ is the so-called chopping number in A/F$^{0.5}$. If the circuit breaker has N interrupters in series, then Eq. (6.14) becomes

$$i_c = \lambda \sqrt{NC_p}. \tag{6.15}$$

Typical values of λ for single interrupters are given in Table 6.2 for the different circuit breaker types. For minimum-oil circuit breakers, the chopping number is a constant independent of arcing time. For air-blast and SF$_6$ circuit breakers, the chopping number increases with increasing arcing time. The chopping number is a characteristic of the circuit breaker and is determined during type testing.

The capacitance in parallel with the circuit breaker (C_p) is given by (Figure 6.9)

$$C_p = C_G + \frac{C_S C_L}{C_S + C_L}, \tag{6.16}$$

where C_G is the circuit breaker grading capacitance or simply stray capacitance and C_S and C_L are the source- and load-side capacitances, respectively. C_S is usually taken as being very much greater than C_L, and Eq. (6.16) reduces to

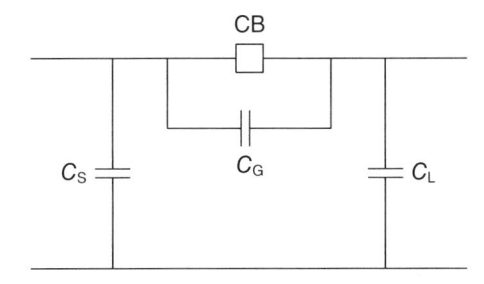

$$C_p = C_G + C_L$$

Figure 6.9 Circuit breaker capacitance circuit.

and Eq. (6.15) to

$$i_c = \lambda\sqrt{N(C_G + C_L)}. \tag{6.17}$$

Substituting for i_c, Eq. (6.7) gives

$$\begin{aligned}k_a &= \sqrt{(k_{in}+K)^2 + \frac{L(1+K)\lambda^2 N(C_G+C_L)}{C_L V^2} - K} \\ &= \sqrt{(k_{in}+K)^2 + L(1+K)\left(\frac{\lambda}{V}\right)^2\left(\frac{C_G}{C_L}+1\right) - K}.\end{aligned} \tag{6.18}$$

Further modification is possible by incorporating the rating of the shunt reactor Q in VA:

$$Q = \sqrt{3} V_L \cdot I,$$

where V_L is the line-to-line voltage and I is the shunt reactor current. Remembering that V is a peak voltage, we can write

$$\begin{aligned}Q &= \sqrt{3}\left(\frac{\sqrt{3}V}{\sqrt{2}}\right)\left(\frac{V}{\sqrt{2}\omega L}\right) \\ &= \frac{1.5}{\omega}\left(\frac{V^2}{L}\right)\end{aligned}$$

or

$$\frac{L}{V^2} = \frac{1.5}{\omega Q},$$

and substitution in Eq. (6.18) gives

$$k_a = \sqrt{(k_{in}+K)^2 + \frac{1.5(1+K)\lambda^2 N}{\omega Q}\left(\frac{C_G}{C_L}+1\right) - K}. \tag{6.19}$$

The use of the equations derived so far is not as complex as it might seem. This will be illustrated in the following sections for the three shunt reactor neutral earthing cases.

6.3 Shunt Reactors with Isolated Neutrals

Shunt reactors with isolated neutrals are applied at system voltages up to 170 kV. Those applied at so-called medium voltages of 52 kV or less are typically on the tertiaries of system transformers. Above 52 kV, the shunt reactors are connected directly to busbars. Each case will be treated separately using actual application examples.

Example 6.1

12 kV 35 Mvar dry-type shunt reactor at 50 Hz.

Assuming $K = 0.5$, $N = 1$ and C_G, being stray capacitance only, to be negligible, Eq. (6.19) becomes

$$k_a = \sqrt{(k_{in}+0.5)^2 + \frac{2.25\lambda^2}{\omega Q}} - 0.5. \qquad (6.20)$$

Assuming an SF$_6$ gas circuit breaker (the alternate use of a vacuum circuit is discussed later), the value of k_{in} is taken as 1.1 pu and λ typically as 0.4×10^4 A/F$^{0.5}$. Substitution in Eq. (6.20) gives

$$k_a = \sqrt{2.56 + 0.003} - 0.5$$
$$= 1.1 \text{ pu.}$$

The second term under the square root sign is insignificant, and k_a is determined by the arc voltage. This is not atypical for medium-voltage shunt applications, but, of course, the user should confirm if this is the case for the actual applications assuming the use of an SF$_6$ gas circuit breaker.

For the shunt reactor circuit, $L = 13$ mH and C_L (for dry-type reactors) is 500 pF. The frequency of the load-side oscillation is

$$f = \frac{1}{2\pi\sqrt{1.5(0.013)(500 \times 10^{-12})}}$$
$$= 50.97 \text{ kHz.}$$

Applying the generalized series RLC circuit case (refer to Section 2.3), we first calculate the value of 1 pu t_g.

$$1 \text{ pu } t_g = \sqrt{1.5(0.013)(500 \times 10^{-12})}$$
$$= 3.12 \text{ μs.}$$

From Eq. (6.6), 1 pu transient voltage is given by

$$1 \text{ pu voltage} = \frac{12\sqrt{2}}{\sqrt{3}}(k_a + K)$$
$$= 9.8(1.1 + 0.5)$$
$$= 15.68 \text{ kV}$$
$$= \text{precharge voltage on } C_L.$$

The TRV across the circuit breaker is given by Eq. (6.12) rearranged as follows to reflect Figure 6.8:

$$k_{rv} = (1+K) + (k_{af} - 1)(k_a + K). \qquad (6.21)$$

The first term on the right-hand side of the equation is the source voltage peak with respect to the shifted neutral and the second term is the load-side voltage peak, also with respect to the shifted neutral and of opposite polarity to the source voltage. Expressing the TRV peak as a voltage V_{pk}, we obtain

$$V_{pk} = 9.8[(1+0.5)+(0.9)(1.1+0.5)]$$
$$= 28.8 \, \text{kV}.$$

The equation for the TRV is given by

$$v(t) = 14.7 - 15.68 \, e^{-0.033 t_g}\left(\cos t_g + 0.033 \sin t_g\right). \quad (6.22)$$

The AC component voltage (14.7 kV), the load-side transient voltage and the resultant TRV according to Eq. (6.22) are shown in Figure 6.10.

To now calculate the rate of rise of recovery voltage (RRRV), the time to peak T is given by

$$T = \pi t_g$$
$$= 9.8 \, \mu s.$$

From Figure 4.15 and $k_{af} = 1.9$, $t_3/T = 0.873$ and $t_3 = 8.55 \, \mu s$,

$$\text{RRRV} = \frac{28.8 + 0.98}{t_3} \quad (\text{refer to Figure 6.10})$$
$$= \frac{29.78}{8.55}$$
$$= 3.48 \, \text{kV}/\mu s.$$

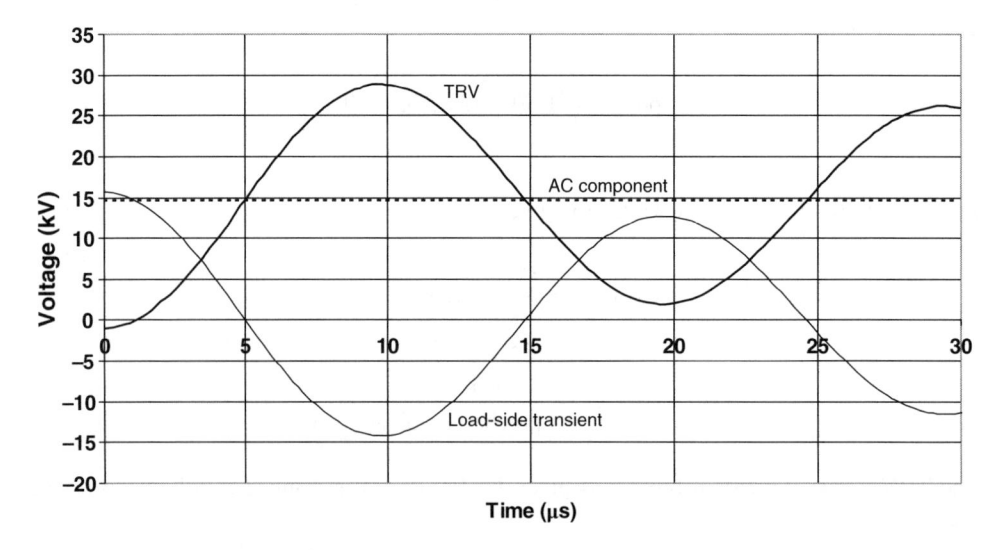

Figure 6.10 TRV for switching out a 12 kV, 35 Mvar shunt reactor with isolated neutral using an SF$_6$ circuit breaker and $C_L = 500 \, \text{pF}$.

Inductive Load Switching

The RRRV value can be reduced by adding a capacitor to earth across the shunt reactor. Assuming that we add 10 000 pF for a total of 10 500 pF, what is the effect on V_{pk} and RRRV? The value of k_{af} can still be taken as 1.9 pu, and we need only to calculate a new value for t_g and RRRV:

$$1 \text{ pu } t_g = \sqrt{1.5(0.013)(10\,500 \times 10^{-12})}$$
$$= 14.3 \text{ }\mu s,$$
$$T = 45 \text{ }\mu s,$$
$$t_3 = 39.3 \text{ }\mu s,$$
$$\text{RRRV} = 0.76 \text{ kV/}\mu s.$$

Equation (6.22) is also valid for this case and is shown in Figure 6.11 compared with the $C_L = 500$ pF case.

For vacuum circuit breakers, k_a is calculated using Eq. (6.7) and taking k_{in} as negligible. For the case of $C_L = 500$ pF and assuming $i_c = 2$ A,

$$k_a = \sqrt{1.5^2 + \frac{0.013(1.5)(2)^2}{(500 \times 10^{-12})(9.8 \times 10^3)^2}} - 0.5$$
$$= 1.47,$$
$$V_{pk} = 9.8[1.5 + 0.9(1.47 + 0.5)]$$
$$= 32 \text{ kV},$$
$$\text{RRRV} = \frac{32 + 4.5}{8.55} = 4.27 \text{ kV/}\mu s.$$

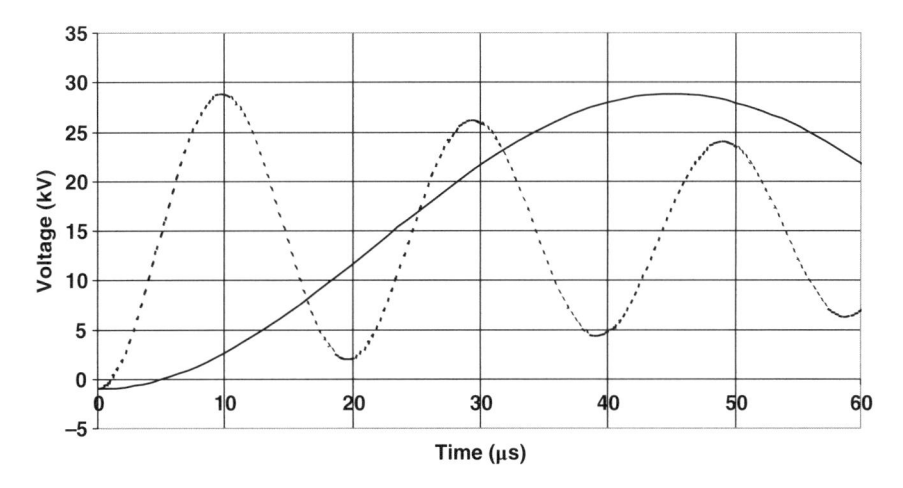

Figure 6.11 TRV for switching out a 12 kV, 35 Mvar shunt reactor with isolated neutral using an SF$_6$ circuit breaker and $C_L = 10\,500$ pF.

The equation for the TRV is

$$v(t) = 14.7 - 19.2\, e^{-0.033 t_g} \left(\cos t_g + 0.033 \sin t_g\right)$$

and comparison with Eq. (6.21) shows that the vacuum circuit breaker is exposed to a higher TRV than the SF_6 gas circuit breaker. This is of, course, because of the added effect of current chopping and associated stored energy. However, the former circuit breakers generally have high RRRV capability, and the addition of capacitors may not be necessary. Each application needs to be judged on its own merit.

Example 6.2

170 kV 35 Mvar oil-filled shunt reactor at 50 Hz.

Shunt reactors with isolated neutrals are in use at transmission voltages up to 170 kV. As the application voltage increases for a fixed shunt reactor rating, the value of the reactor inductance increases, and the frequency of the load-side transient decreases.

We will consider again an oil-filled 35 Mvar shunt reactor but now rated at 170 kV and switched by a single interrupter 170 kV SF_6 gas circuit breaker with no grading capacitor. The following parameters apply for the TRV calculation:

- Shunt reactor: $L = 2.63$ H and $C_L = 2000$ pF.
- Circuit breaker: $C_G = 0$, $k_{in} = 1.05$, $N = 1$, $\lambda = 4 \times 10^4$ A/F$^{0.5}$.
- Load circuit: $K = 0.5$, $k_{af} = 1.9$, $d_s = 0.033$.

Applying Eq. (6.19),

$$k_a = \sqrt{(1.05+0.5)^2 + \frac{(1.5)^2 (4 \times 10^4)^2}{314(35 \times 10^6)}} - 0.5$$
$$= 1.15,$$

$$1 \text{ pu source voltage} = \frac{170\sqrt{2}}{\sqrt{3}} = 138.8 \text{ kV},$$

$$1 \text{ pu transient voltage} = 138.8(1.15 + 0.5)$$
$$= 229 \text{ kV},$$

$$V_{pk} = 138.8[1.5 + 0.9(1.15 + 0.5)]$$
$$= 414.3 \text{ kV},$$

$$t_g = \sqrt{1.5(2.63)(2000 \times 10^{-12})}$$
$$= 88.2 \text{ μs}.$$

The equation for the TRV is

$$v(t) = 208 - 229\, e^{-0.033 t_g} \left(\cos t_g + 0.033 \sin t_g\right)$$

and shown in Figure 6.12.

Inductive Load Switching

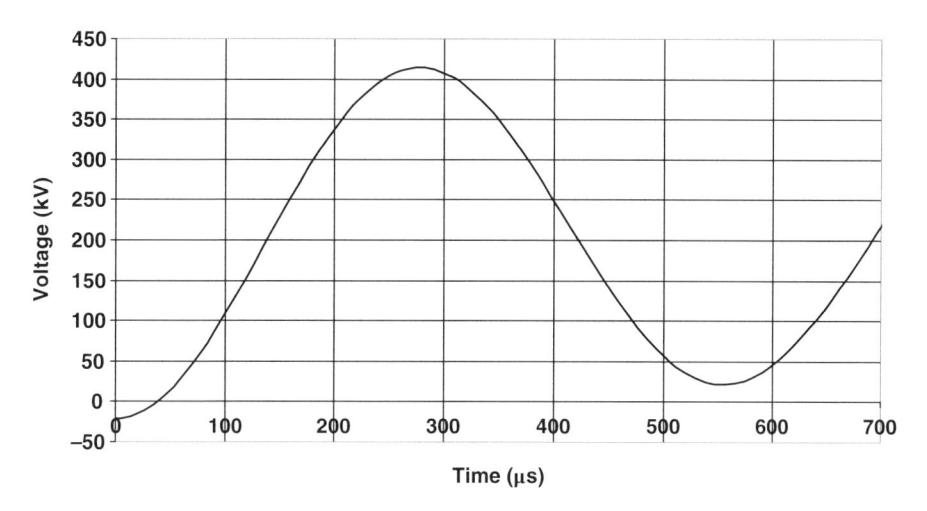

Figure 6.12 TRV for switching out a 170 kV, 35 Mvar shunt reactor with isolated neutral using an SF_6 circuit breaker.

To calculate the RRRV value,

$$T = \pi t_g = 277 \, \mu s,$$
$$t_3 = 277 \times 0.873 = 242 \, \mu s,$$
$$\text{RRRV} = \frac{414.3 + 21}{242} = 1.8 \, \text{kV}/\mu s.$$

As illustrated in Examples 1 and 2, the fact that the shunt reactor neutral is isolated contributes 0.95 pu to the peak TRV value.

6.4 Shunt Reactors with Neutral Reactor Earthed Neutrals

Shunt reactors with neutral reactor earthed neutrals are most common on extra-high-voltage (EHV) systems, that is, at system voltages of 300 kV or greater, to enable single pole fault clearing (refer to Figure 6.1). The purpose of the neutral is to ensure that the secondary arc is extinguished before reclosing on the faulted phase.

We will take the case of 516 kV, 135 Mvar shunt reactor with a 1000 Ω neutral reactor at 50 Hz and switched by a two-interrupter SF_6 circuit breaker. The following parameters apply for the TRV calculation:

- Shunt reactor: rated current 151 A, $L = 6.28$ H and $C_L = 2000$ pF.
- Neutral reactor: $L_n = 3.18$ H.
- Circuit breaker: grading capacitance 1600 pF per interrupter, $C_G = 1600/2 = 800$ pF, $k_{in} = 1.05$, $N = 2$, $\lambda = 10 \times 10^4$ A/F$^{0.5}$.
- Load circuit: $K = 1/[2 + (6.28/3.18)] = 0.25$, $k_{af} = 0.9$, $d_s = 0.033$.

Applying Eq. (6.19),

$$k_a = \sqrt{(1.05+0.25)^2 + \frac{(1.5)(1.25)(10 \times 10^4)^2 (2)}{(314)(135 \times 10^6)} \left(\frac{800}{2000}+1\right)} - 0.25$$
$$= 1.46 \text{ pu},$$

$$1 \text{ pu source voltage} = \frac{516\sqrt{2}}{\sqrt{3}} = 412.3 \text{ kV},$$

$$1 \text{ pu transient voltage} = 412.3(1.46+0.25)$$
$$= 705 \text{ kV},$$

$$V_{pk} = 412.3[1.25 + 0.9(1.46+0.25)]$$
$$= 1150 \text{ kV},$$
$$t_g = \sqrt{1.25(6.28)(2000 \times 10^{-12})}$$
$$= 125.3 \text{ μs}.$$

The equation for the TRV is

$$v(t) = 515 - 705 \, e^{-0.033 t_g} \left(\cos t_g + 0.033 \sin t_g\right). \tag{6.23}$$

To calculate the RRRV value,

$$T = \pi t_g = 393.6 \text{ μs},$$
$$t_3 = 393.6 \times 0.873 = 343.6 \text{ μs},$$
$$\text{RRRV} = \frac{1150 + 190}{343.6} = 3.9 \text{ kV/μs}.$$

6.5 Shunt Reactors with Earthed Neutrals

Shunt reactors with earthed neutrals are the most common of all shunt reactor installations. Taking the same case as for Section 6.4, that is, setting $K = 0$, the TRV is calculated as follows:

$$k_a = \sqrt{1.05^2 + \frac{(1.5)(10 \times 10^4)^2 (2)}{(314)(135 \times 10^6)} \left(\frac{800}{2000}+1\right)}$$
$$= 1.446 \text{ pu},$$

$$1 \text{ pu transient voltage} = 412.3(1.446)$$
$$= 596 \text{ kV},$$

$$V_{pk} = 412.3[1 + 0.9(1.446)]$$
$$= 949 \text{ kV},$$
$$t_g = \sqrt{(6.28)(2000 \times 10^{-12})}$$
$$= 112 \text{ μs}.$$

Inductive Load Switching

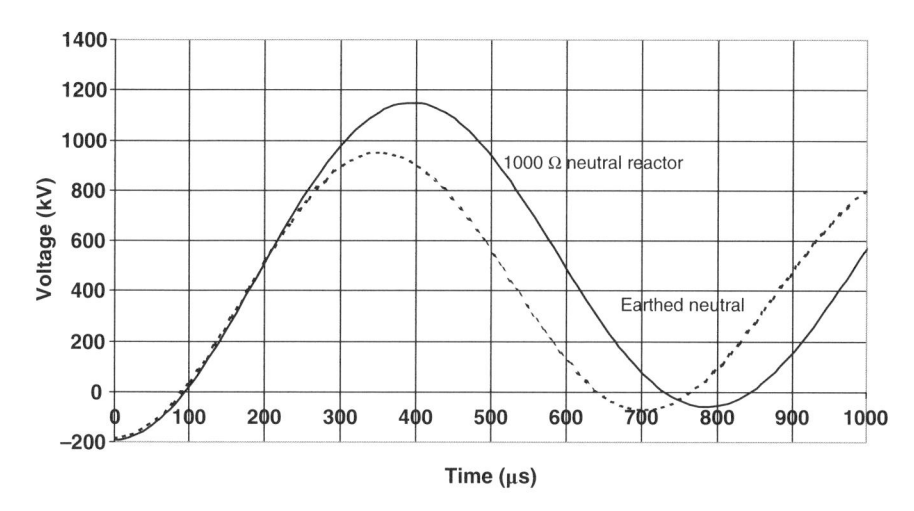

Figure 6.13 TRVs for switching out a 516 kV, 135 Mvar shunt reactor with a 1000 Ω neutral reactor and with an earthed neutral.

The equation for the TRV is

$$v(t) = 412.3 - 596\, e^{-0.033 t_g} \left(\cos t_g + 0.033 \sin t_g\right). \tag{6.24}$$

To calculate the RRRV value,

$$T = \pi t_g = 352\,\mu s,$$
$$t_3 = 352 \times 0.873 = 307\,\mu s,$$
$$\text{RRRV} = \frac{949 + 183.7}{307} = 3.7\,\text{kV}/\mu s.$$

Equations (6.23) and (6.24) are shown in Figure 6.13. The increased severity of the neutral reactor case is obvious. In practice, it is common to bypass the neutral reactor by means of a single-pole disconnector before switching out the shunt reactor.

6.6 Re-ignitions

Re-ignitions will occur if the arcing time—the time between circuit breaker contact parting and the first occurring current zero crossing—is too short. The re-ignition circuit is shown in Figure 6.14 and involves the source- and load-side capacitances C_S and C_L and the inductance of the interconnections L_b. This circuit is known as the second parallel oscillation circuit. Re-ignition overvoltage oscillations are circuit dependent and are not characteristic in any way of the circuit breaker. The oscillation frequency is determined by the L_b–C_S–C_L series circuit and is in the range of 50 kHz to 1 MHz.

Re-ignition at the TRV peak will give the highest re-ignition voltage as shown schematically in Figure 6.15. The starting point of the re-ignition oscillation is at the TRV peak, the axis of oscillation is 1 pu system voltage and the oscillation excursion (k_s) and peak value to earth (k_p)

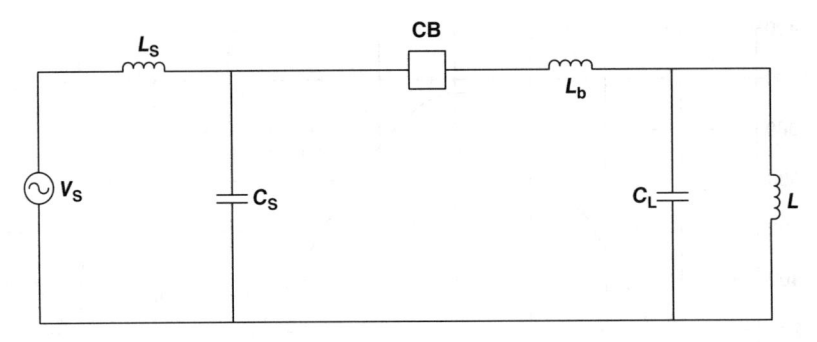

Figure 6.14 Second parallel re-ignition oscillatory circuit.

are given in Eqs. (6.24) and (6.25), respectively:

$$k_s = (1+\beta)k_{rv}, \tag{6.25}$$

$$k_p = 1 + \beta k_{rv}, \tag{6.26}$$

where β is the damping factor of the re-ignition circuit. Based on experience, β is usually taken as 0.5 pu.

Re-ignitions in themselves stress the circuit breaker (as discussed in Section 6.7), and the excursion imposes a stress on the shunt reactor windings. If the current associated with the re-ignition is not interrupted by the circuit breaker, the system becomes involved, and another loop of power frequency current will flow with the interruption repeated at the next current zero crossing.

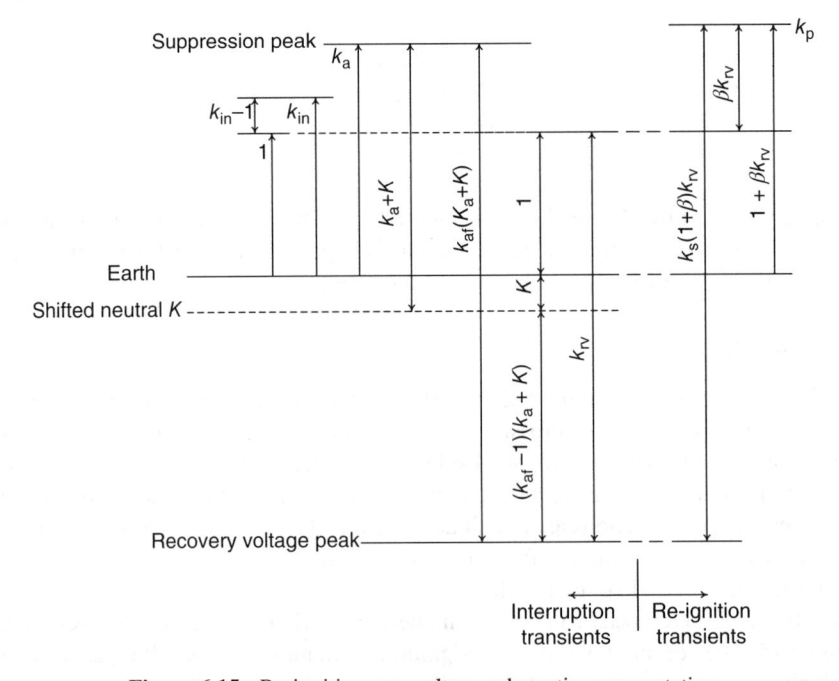

Figure 6.15 Re-ignition overvoltage schematic representation.

6.7 Unloaded Transformer Switching

Similar to shunt reactors, an unloaded transformer can be represented by an underdamped series RLC circuit. However, the circuit is quite heavily damped at a frequency in the order of a few hundred hertz. The peak of the load-side oscillation does not coincide in time with the source voltage, and the latter voltage can be considered as both the "transient" and power frequency recovery voltage.

The magnetizing current for modern low-loss transformers is less than 1 A and presents no difficulty for circuit breakers to interrupt. For air-break disconnectors, current interruption involves extinction of free burning arc. At 1 A or less, such arcs do not exhibit thermal effects, and arc extinction is basically dependent on achieving a gap spacing sufficient to withstand the source voltage.

6.8 Discussion

Shunt reactor switching is an onerous duty for circuit breakers. No standard ratings can be applied, and each application has to be considered based on the actual shunt reactor type and rating, neutral earthing and possible additional load-side capacitors.

For circuit breakers applied particularly at 72.5 kV and above, experience has shown that re-ignitions are to be avoided. Ideally, re-ignitions should occur between the circuit breaker arcing contacts, but this has not proven to be the case for the low currents associated with shunt reactor switching. In fact, re-ignitions have been known to occur between an arcing and a main contact and puncturing the nozzle. To avoid re-ignitions, controlled switching has become standard practice. For example, if a circuit breaker has a minimum arcing time of 3 ms, as determined in a type test, then a controller is used to achieve a contact parting of 6 ms before zero crossings on each phase, thus avoiding re-ignitions.

The chopping number is a characteristic of the circuit breaker and is statistical as also determined in a type test. For oil circuit breakers, now largely replaced, the chopping numbers do not vary with arcing time. For gas circuit breakers, however, the chopping numbers increase approximately linearly with arcing time from the value at the minimum arcing.

In Eq. (6.7), readers would have noted that the second term under the square root sign is proportional to the inductance L; in Eq. (6.9), the same term is inversely proportional to Q. This means that the higher the value of L or the lower the value of Q, the higher the value of k_a and the TRV imposed on the circuit breaker (refer to Section 8.4). Therefore, the lower the current, the more onerous the duty, and type testing is performed at the minimum current required by the relevant IEC standard (see Bibliography) or that of the actual application if that current is even lower.

6.9 Summary

Shunt reactor switching is a difficult duty for circuit breakers. A number of variations exist dependent on the earthing of the shunt reactor neutrals and on the type of circuit breaker. With respect to the circuit breaker TRVs, dependences can be summarized as follows:

- Current chopping results in stored energy in the shunt reactor that contributes to the magnitude of the TRV. This is particularly the case for vacuum circuit breakers applied at 52 kV and below and for SF_6 gas circuit breakers at 52 kV and above.

Table 6.3 Shunt reactor switching equation summary.

Calculation	Equation	Comment
Suppression peak overvoltage (general approach)	$k_a = \sqrt{(k_{in}+K)^2 + \dfrac{L(1+K)i_c^2}{C_L V^2} - K}$	Use for all circuit breaker types
Suppression peak overvoltage (chopping number approach)	$k_a = \sqrt{(k_{in}+K)^2 + \dfrac{1.5(1+K)\lambda^2 N}{\omega Q}\left(\dfrac{C_G}{C_L}+1\right) - K}$	Use for all circuit breaker types except vacuum circuit breakers
TRV peak	$k_{rv} = (1+K) + (k_{af}-1)(k_a+K)$	Use for all circuit breaker types
Re-ignition overvoltage excursion	$k_s = (1+\beta)k_{rv}$	
Re-ignition overvoltage peak to earth	$k_p = 1 + \beta k_{rv}$	

- For shunt reactors with isolated neutrals or earthed through neutral reactors, the first pole TRV is increased by a factor approaching twice the neutral shift.
- For vacuum circuit breakers, TRV values can be calculated using the circuit elements and the chopped current level.
- For SF_6 gas circuit breakers, TRV values can be calculated using the chopping number approach. Chopping numbers are statistical and increase with arcing time.

The equations for TRV and re-ignition overvoltage calculation are summarized in Table 6.3.

Bibliography

1. CIGRE (1995) Interruption of Small Inductive Currents. CIGRE Technical Brochure 50.
2. IEC62271-306 (2012) *High-Voltage Switchgear and Controlgear–Part 306: Guide to IEC 62271–100, IEC 62271-1 and Other IEC Standards Related to Alternating Current Circuit-Breakers*, International Electrotechnical Commission.
3. IEC62271-110 (2012) *High-Voltage Switchgear and Controlgear–Part 110: Inductive Load Switching*, International Electrotechnical Commission.

7

Capacitive Load Switching

7.1 General

The capacitive load switching cases to be considered are the switching of shunt capacitor banks, unloaded transmission lines and unloaded cables. Capacitive load currents are 90° leading, and current interruption leaves a DC voltage on the load circuit. The recovery voltage across the circuit breaker will thus be the difference between the source AC voltage and the load-side DC voltage as shown in Figure 7.1 for banks with earthed neutrals.

Similar to inductive load switching, there are a number of load circuit configurations of interest, and these will be addressed as we cover the various cases. However, capacitive current switching differs from inductive load switching duties in that energization inrush currents are a major consideration. An extension to this are outrush currents, that is, discharge of a capacitor bank through a circuit breaker into a fault. Normal practice is to apply some form of mitigation to limit inrush current magnitudes and frequencies to values specified in IEC 62271-100.

For each of the capacitive load circuit configurations, we will discuss both energization and de-energization and associated current or voltage transients and their mitigation as applicable.

7.2 Shunt Capacitor Banks

Shunt capacitor banks are three-phase lumped capacitive circuit elements with either earthed or unearthed neutral points. Two cases need to be considered, single capacitor banks in which the energization is relative to the system and back-to-back capacitor banks (i.e. multiple banks connected to the same bus but switched individually) in which the energization is relative to the banks already energized.

7.2.1 Energization

The general circuit for shunt capacitor bank energization analysis is shown in Figure 7.2, which readers will recognize as being identical to Figure 2.1a. Random closing of the circuit breaker applies a virtual step voltage V to the bank, and assuming no prior voltage on the bank, the bank must charge almost instantaneously to this voltage. Obviously, the value of V is dependent on the point on the voltage half-wave at which prestrike or contact touch occurs in the circuit breaker. The maximum value of V is the peak value of the applied voltage.

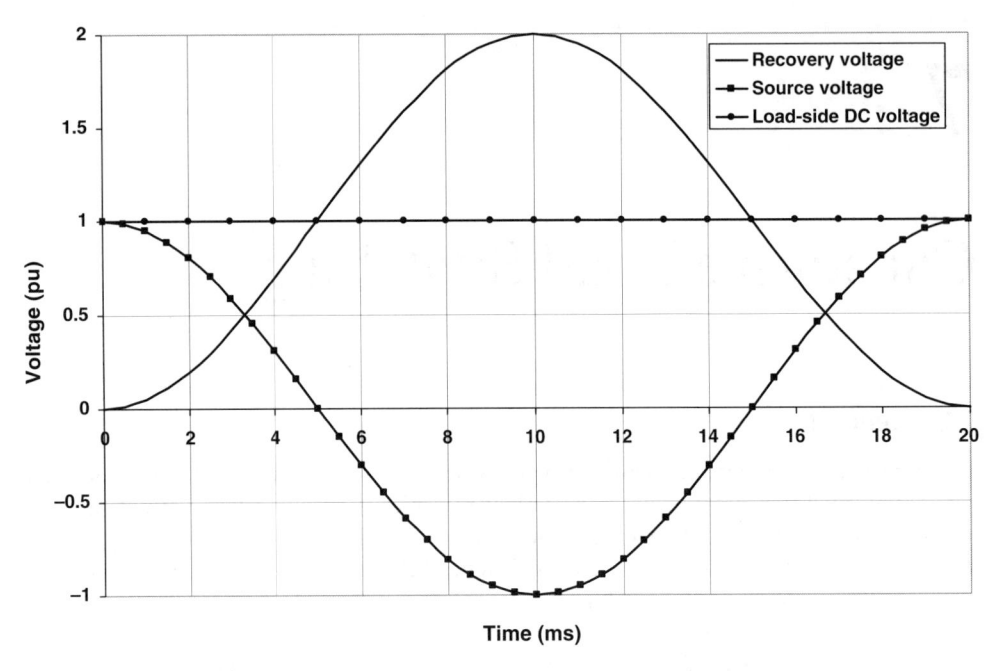

Figure 7.1 Capacitive current interruption recovery voltage.

7.2.1.1 Inrush Current

The capacitor bank is charged by the so-called inrush current from the system having a frequency usually much greater than the power frequency.[1] In Figure 7.2, C is the capacitance of the bank in farads, L is the source impedance in henries and R represents the losses in the circuit in ohms. We derived the equations for the current in Section 2.2 and from Table 2.1:

Overdamped case: $d_s > 1$ and $R > R_C$.

$$i(t) = \frac{V}{Z\sqrt{d_s^2 - 1}} e^{-d_s t_g} \sinh \sqrt{d_s^2 - 1}\, t_g. \tag{7.1}$$

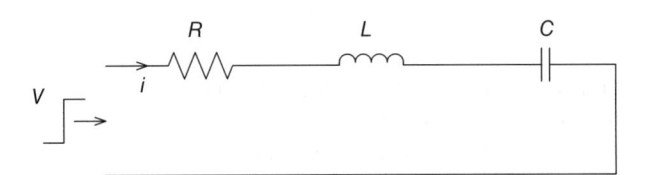

Figure 7.2 Circuit for capacitive inrush current calculation.

---

[1] The inrush current causes a voltage dip at the bank location. The reason for this is that the source impedance is higher for the higher inrush current frequency, and the impedance of the bank is correspondingly lower. This is a power quality issue, and mitigation is usually necessary to limit the voltage dip.

Capacitive Load Switching

Critically damped case: $d_s = 1$ and $R = R_C$.

$$i(t) = \frac{V}{Z} t_g e^{-t_g}. \tag{7.2}$$

Underdamped case: $d_s < 1$ and $R < R_C$.

$$i(t) = \frac{V}{Z\sqrt{1-d_s^2}} e^{-d_s t_g} \sin\sqrt{1-d_s^2}\, t_g, \tag{7.3}$$

all where $R_C = 2\sqrt{\frac{L}{C}}$ and $d_s = R/R_C$.

In general, $R \ll R_C$ and Eq. (7.3) can be written in terms of real time as

$$i(t) = \frac{V}{Z} e^{-\alpha t} \sin \omega t, \tag{7.4}$$

where $\alpha = R/2L$ and $\omega = 1/\sqrt{LC}$. If damping is ignored, the peak value of the inrush current is V/Z. This is a useful concept that readers should keep in mind when dealing with prestriking, re-ignitions and re-striking.

The circuit for the back-to-back switching case is shown in Figure 7.3. C_2 is initially uncharged, and the switch is closed at a voltage V on C_1. L_{eff} is the effective inductance of connecting leads and buswork. Applying Kirchoff's voltage law, we can write

$$\frac{1}{C_1}\int i\, dt + Ri + L_{\text{eff}} \frac{di}{dt} + \frac{1}{C_2}\int i\, dt = 0.$$

Differentiating across,

$$L_{\text{eff}} \frac{d^2 i}{dt^2} + R \frac{di}{dt} + \left(\frac{1}{C_1} + \frac{1}{C_2}\right) i = 0,$$

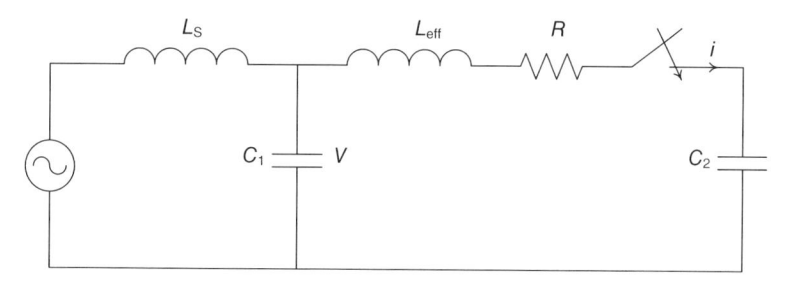

Figure 7.3 Circuit for back-to-back capacitive inrush current calculation.

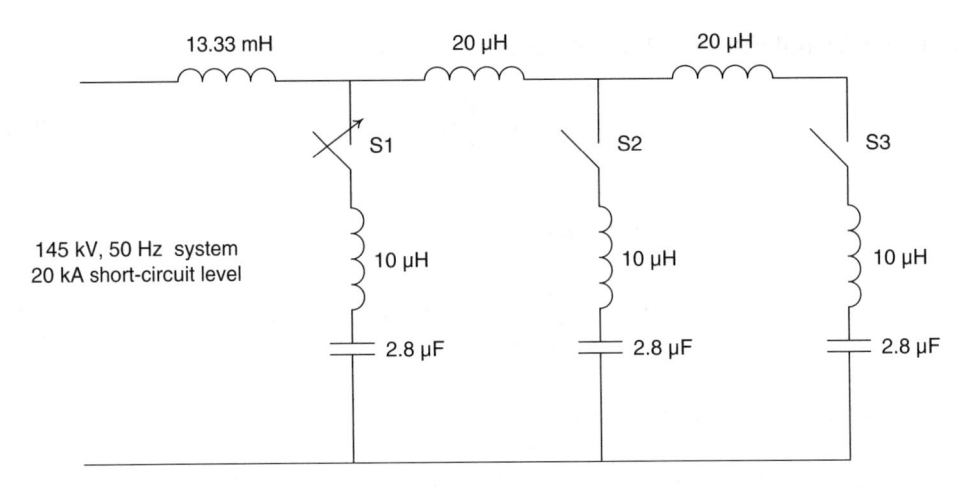

Figure 7.4 Basic circuit for single and back-to-back capacitor switching example.

giving

$$\frac{d^2 i}{dt^2} + \frac{R}{L_{\text{eff}}} \frac{di}{dt} + \frac{1}{L_{\text{eff}} C_{\text{eff}}} i = 0, \qquad (7.5)$$

where $C_{\text{eff}} = C_1 C_2/(C_1 + C_2)$.

Equation (7.5) has the same format as Eq. (2.2), and the same solutions for the various damping cases are equally applicable using the appropriate values for R, L and C. Likewise, the generic damping curves of Figure 2.2 also apply. To derive the values for L_{eff} and C_{eff}, an acceptable approximation is to ignore the inductances and calculate C_{eff} and then ignore the capacitances and calculate L_{eff}.

By way of example, consider the circuit shown in Figure 7.4. Three switchable earthed neutral shunt capacitor banks with $C = 2.8\,\mu\text{F}$ and connection inductances as shown are applied on a 145 kV, 50 Hz system with a short-circuit level of 20 kA, giving a source inductance of 13.34 mH. The peak system voltage to earth is 118.4 kV.

Case 1: Single shunt capacitor switching.

Switches S1, S2 and S3 are initially open, and switch S1 is closed to energize the first bank. The switches can be either circuit breakers or load interrupter switch devices.

Surge impedance: $Z = \sqrt{\dfrac{13.34 \times 10^{-3}}{2.8 \times 10^{-6}}} = 69\,\Omega$.

Inrush current peak (no damping): $I_{\text{pk}} = \dfrac{118.4}{69} = 1.7\,\text{kA}$.

Inrush current frequency: $f = 823\,\text{Hz}$.

Critical resistance: $R_C = 2 \times 69 = 138\,\Omega$.

Degree of damping: assume $d_s = 0.1 \rightarrow R = 13.8\,\Omega$.

Damping coefficient: $\alpha = \dfrac{R}{2L} = 517$.

1 pu t_g: $\sqrt{13.34 \times 10^{-3} \times 2.8 \times 10^{-6}} = 193\,\mu\text{s}$.

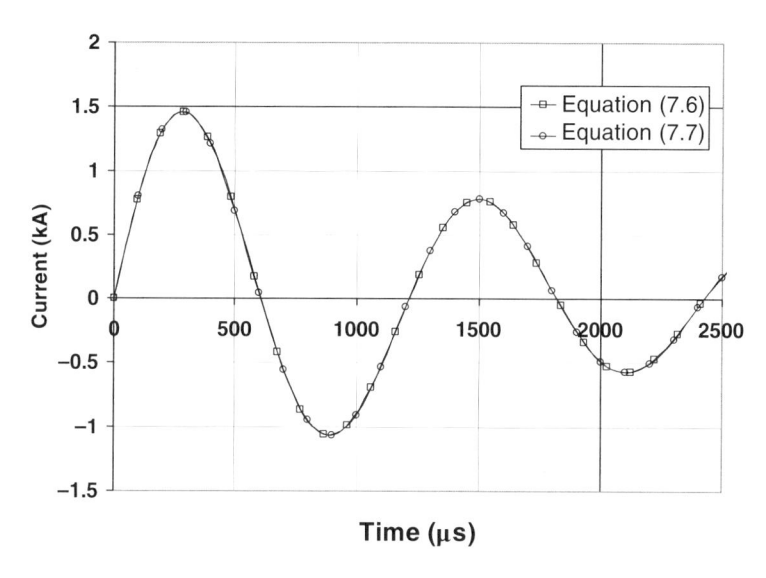

Figure 7.5 Inrush current for single capacitor bank switching example.

Taking $\sqrt{1-d_s^2} \approx 1$, from Eq. (7.3) we can write the equation for the current in kA:

$$i = 1.7\,e^{-0.1 t_g} \sin t_g \quad (7.6)$$

or from Eq. (7.4)

$$i = 1.7 e^{-517 t} \sin \omega t, \quad (7.7)$$

where $\omega = 1/\sqrt{LC} = 5180$.

Equations (7.6) and (7.7) are plotted in Figure 7.5; the difference is clearly not discernible.
Case 2: Back-to-back shunt capacitor bank switching.
Switches S1 and S2 are already closed, and switch S3 is closed to energize the third bank. The derivation of the effective circuit is shown in Figure 7.6.

Surge impedance: $Z = \sqrt{\dfrac{37.5 \times 10^{-6}}{1.866 \times 10^{-6}}} = 4.48\,\Omega$.
Inrush current peak (no damping): $I_{pk} = \dfrac{118.4}{4.48} = 26.4\,\text{kA}$.
Inrush current frequency: $f = 19\,\text{kHz}$.
Critical resistance: $R_C = 2 \times 4.48 = 8.96\,\Omega$.
Degree of damping: assume $d_s = 0.1 \rightarrow R = 0.896\,\Omega$.
Damping coefficient: $\alpha = 11947$.
1 pu t_g: 8.36 µs.

Again, taking $\sqrt{1-d_s^2} \approx 1$, from Eq. (7.3) we can write

$$i = 26.4\,e^{-0.1 t_g} \sin t_g. \quad (7.8)$$

Equation (7.8) is plotted in Figure 7.7. Readers are encouraged to verify that the use of Eq. (7.4) will give the same result.

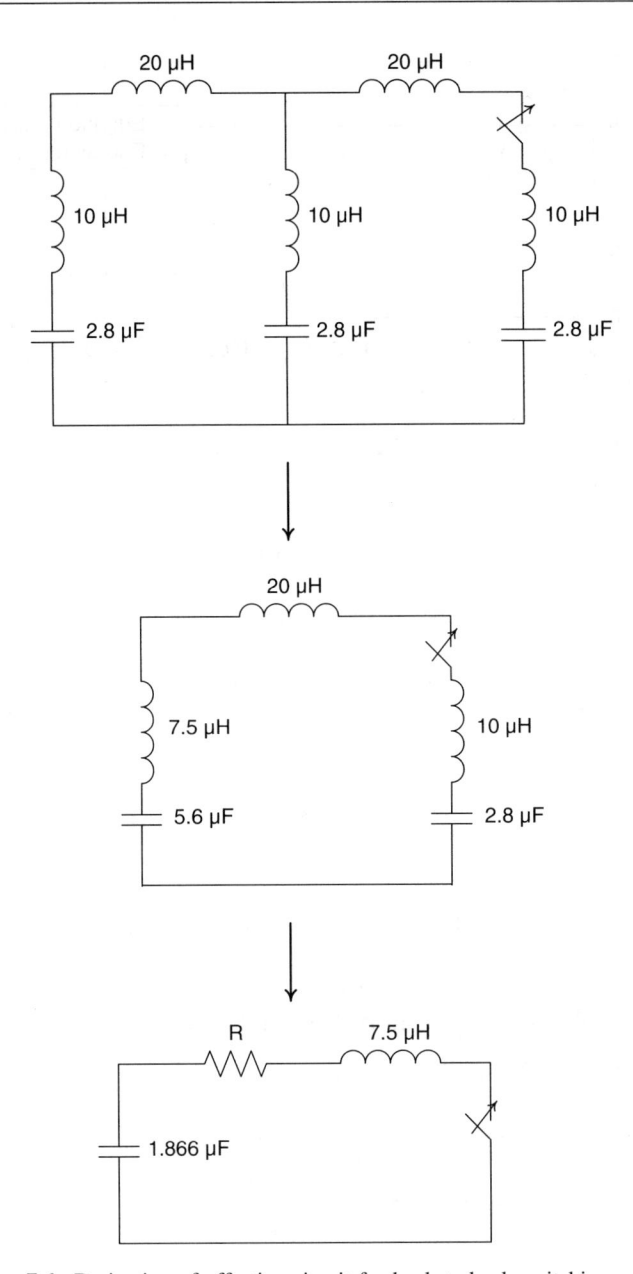

Figure 7.6 Derivation of effective circuit for back-to-back switching example.

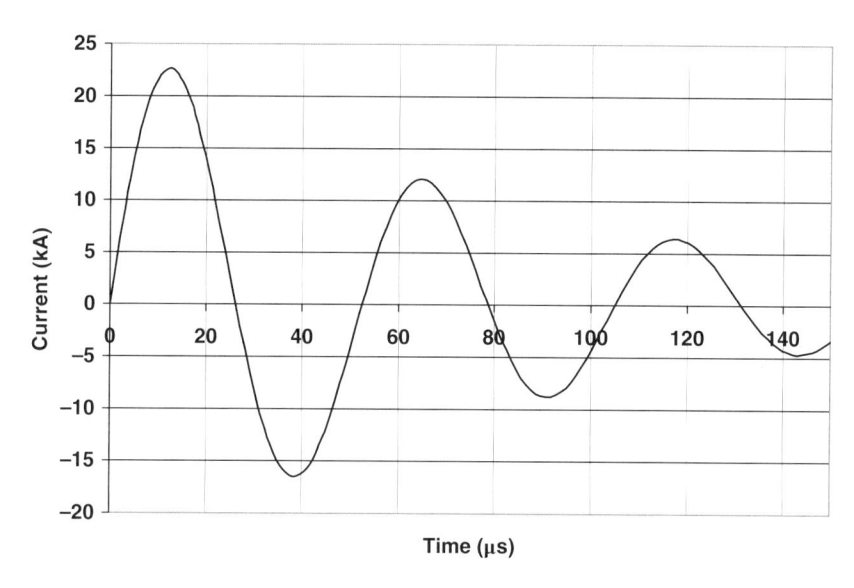

Figure 7.7 Inrush current for back-to-back capacitor bank switching example.

IEC 62271-100 requires that the inrush current peak and frequency for back-to-back switching be limited to 20 kA and 4250 Hz,[2] respectively. These values are both exceeded in the Case 2 example, and some form of limitation is required. Inrush current limitation is discussed in Section 7.2.1.2.

For single capacitor bank energization, the earthed neutral case gives the most onerous inrush current values. With reference to Figure 7.2, V is the peak phase voltage to earth, and we can write the following:

Frequency: $\dfrac{1}{2\pi\sqrt{LC}}$.

Critical resistance: $R_C = 2\sqrt{\dfrac{L}{C}} = 2Z$.

Inrush current peak (no damping): $\dfrac{V}{Z}$.

Degree of damping: $d_s = \dfrac{R}{R_C}$.

The unearthed neutral case is shown in Figure 7.8. The applied voltage is now the line voltage V_L equal to $\sqrt{3}V$. Prestrike or contact touch in two phases is required for the inrush current to occur, and we can write the following:

Frequency: $\dfrac{1}{2\pi\sqrt{2L\cdot(C/2)}} = \dfrac{1}{2\pi\sqrt{LC}}$.

Critical resistance: $R_{Cu} = 4\sqrt{\dfrac{L}{C}} = 4Z = 2R_C$.

---

[2] These values originally applied to bulk-oil type circuit breakers and have persisted in the IEEE and IEC circuit breaker standards ever since. Modern circuit breakers have higher capabilities in this regard; for example, series capacitor bypass switches (actually circuit breakers in practice) have rated bypass making current peak values of 63, 100 and 125 kA at 500 and 1000 Hz. In practice, values of 100 kA at 2000 Hz are common.

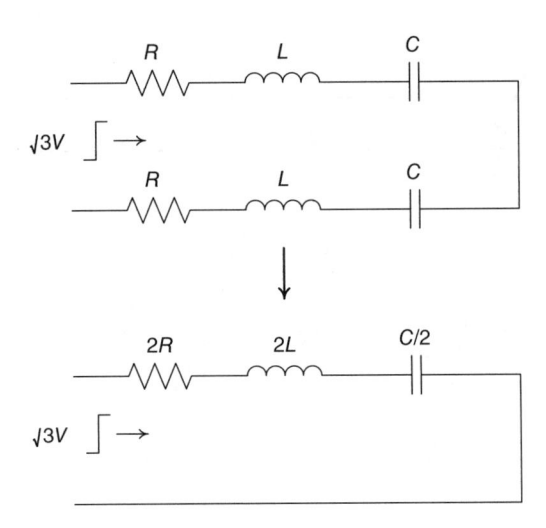

Figure 7.8 Circuit for inrush current calculation for a single capacitor bank with unearthed neutral.

Surge impedance: $Z_u = 2\sqrt{\dfrac{L}{C}} = 2Z$.

Inrush current peak (no damping): $\dfrac{\sqrt{3}V}{2Z} = 0.866\dfrac{V}{Z}$.

Degree of damping: $d_{su} = \dfrac{2R}{R_{Cu}} = \dfrac{R}{R_C}$.

This analysis shows that for the unearthed neutral case compared with the earthed neutral case:

- Frequency and degree of damping: remain unchanged.
- Critical resistance and surge impedance double.
- Inrush current peak is lower.

Readers can determine if the same results are applicable for the back-to-back case or not.

7.2.1.2 Limiting Inrush Current

With reference to Figure 2.2 and Eqs. (7.1)–(7.3), the inrush current and frequency can be influenced by changing the value of d_s ($= R/R_C$) and Z. The choice is to add inductance to increase the value of L_{eff} or to add resistance to increase the value of R. The effects for each are as follows:

- *Adding inductance*: The value of Z increases reducing the undamped peak value V/Z, and the frequency will decrease; the value of d_s will decrease because of the increase in R_C.
- *Adding resistance*: The value of d_s increases, thus reducing the peak value; if the circuit is underdamped, the frequency is not influenced, but by choosing an added resistance value to achieve $d_s = 1$ or greater, the oscillation will become aperiodic.

Taking Case 2 (Figure 7.6) from Section 7.2.1.1, adding a series inductance of 1 mH in each branch will give a circuit with C_{eff} unchanged, and L_{eff} becomes 1.5 mH (bus and connection inductances now insignificant). Again, taking $d_s = 0.1$, the equation for the current is

$$i = 4.17\, e^{-0.1 t_g} \sin t_g. \tag{7.9}$$

Capacitive Load Switching

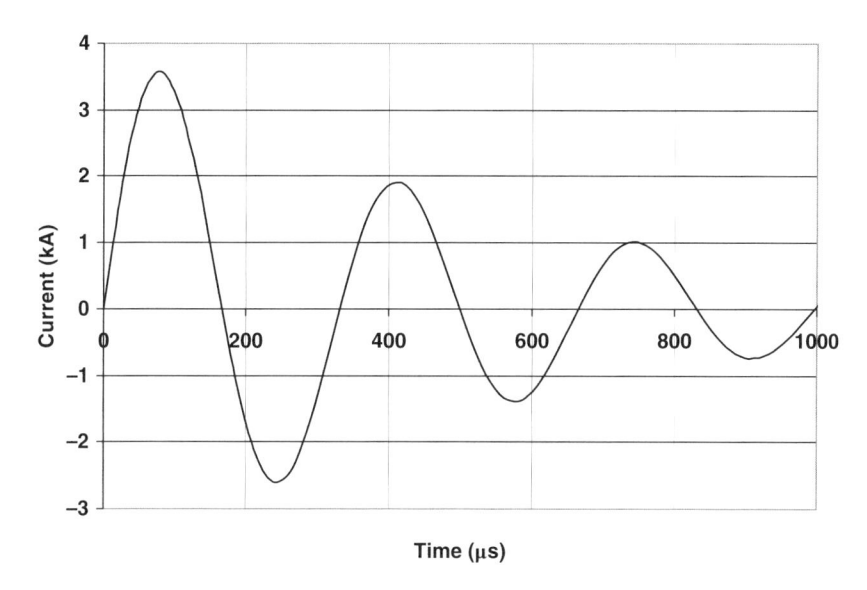

Figure 7.9 Back-to-back switching inrush current example with 1 mH series reactor added in each branch.

Equation (7.9) is plotted in Figure 7.9, and by comparison with Figure 7.7, the current peak and frequency values are now within the limits set by the IEC standard.

To consider the added resistance effect, taking $R = R_C$ to achieve critical damping gives the following equation for the current:

$$i = 26.4 t_g \, e^{-t_g}. \tag{7.10}$$

Equation (7.10) is plotted in Figure 7.10 together with Eq. (7.8) for comparison. Again, the requirements of the IEC standard are met. Achieving this obviously involves the use of a device with appropriately chosen closing resistors.

Another option to limit inrush current is to prevent its occurrence by using controlled switching. This is achieved by means of a controller that aims to cause contact touch close to a voltage zero crossing as detailed in IEC 62271-302.

The use of added inductance, so-called series reactors or current limiting reactors (CLRs), is common and at the discretion of the user. Added resistance is at the discretion of both the user and the switching device supplier. Circuit breakers with closing resistors (also referred to as pre-insertion resistors) are not commonly used for this purpose. However, load break devices with closing resistors are available on the market. Controlled switching is a choice of the user, and the circuit breaker must have the repeatable mechanical closing time accuracy for the application.

Current limiting reactors are the most common means applied to limit inrush current magnitudes. If the reactors are installed on the source side of the shunt capacitor bank, a fault between the reactor and the bank will expose the capacitor bank circuit breaker to the transient recovery voltage (TRV) associated with the reactor. To show this effect and its calculation, we will use the same case as in Section 4.5 except that now the 10 Ω reactor and 12 nF capacitor are on the load side of the circuit breaker.

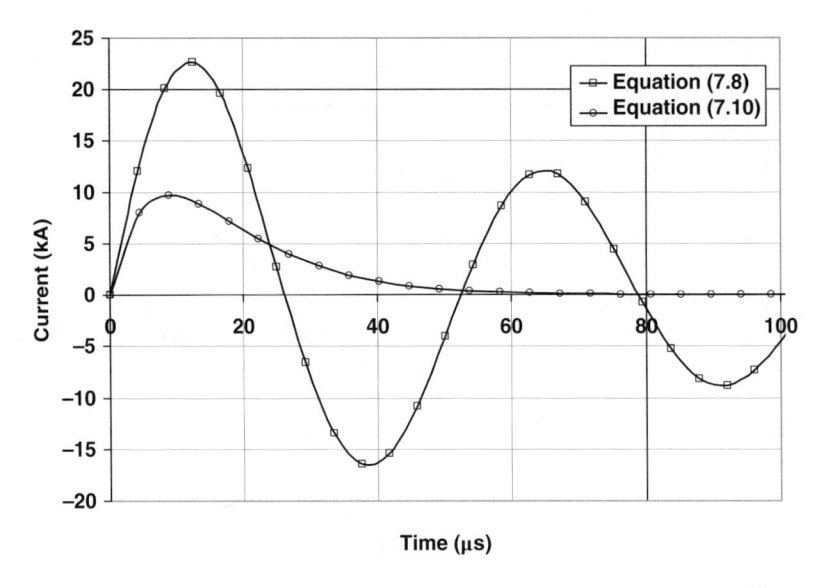

Figure 7.10 Back-to-back switching inrush current with circuit resistance equal to the critical resistance.

The TRVs on either side will oscillate independently, each starting at 172.6 kV, which is the voltage drop across the reactor at current interruption. The source-side TRV v_s is an overdamped oscillation that will recover to the source voltage and is given by

$$v_s = 66.4\left[1 - e^{-3.27 t_{g1}} \left(\cosh 3.1\, t_{g1} + 1.05 \sinh 3.1\, t_{g1}\right)\right],$$

where 1 pu $t_{g1} = 13\,\mu s$.

The load side is represented by a series underdamped oscillatory circuit with the capacitor precharged to 172.6 kV. The load-side oscillation is given by (refer to Section 4.5)

$$v_l = 172.6\, e^{-0.033 t_{g2}} \left(\cos t_{g2} + 0.033 \sin t_{g2}\right),$$

where 1 pu $t_{g2} = 22.29\,\mu s$.

The TRV across the circuit breaker is the difference between the above two oscillations. Figure 7.11 shows the source- and load-side oscillations and the circuit breaker TRV compared with the IEC T60 and T30 test duty TRVs.

An examination of the circuit breaker TRVs calculated in Section 4.5 (Figure 4.28) and in this case (Figure 7.11) shows that they are, in fact, equal. This means that the TRVs as calculated in both cases must have the same equation.

Simplifying Eq. (4.16) gives

$$v_{SR} = 239 - 66.4\, e^{-3.27\, t_{g1}} \left(\cosh 3.1\, t_{g1} + 1.05 \sinh 3.1\, t_{g1}\right) \\ - 172.6\, e^{-0.033\, t_{g2}} \left(\cos t_{g2} + 0.033 \sin t_{g2}\right).$$

Capacitive Load Switching

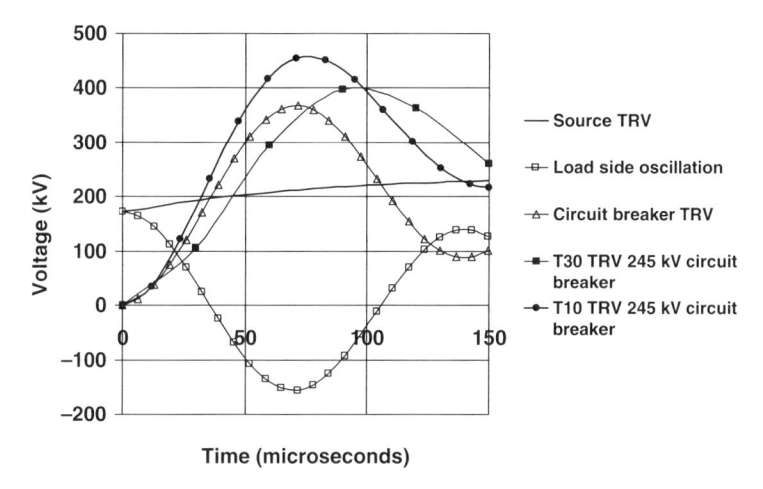

Figure 7.11 TRVs for three-phase-to-earth fault between CLR and shunt capacitor bank.

In turn, from above, the equation for load-side reactor case is

$$\begin{aligned}
v &= 172.6 + v_s - v_l \\
&= 172.6 + 66.4\left[1 - e^{-3.27\,t_{g1}}\left(\cosh 3.1\,t_{g1} + 1.05 \sinh 3.1\,t_{g1}\right)\right] \\
&\quad - 172.6\, e^{-0.033\,t_{g2}}\left(\cos t_{g2} + 0.033 \sin t_{g2}\right) \\
&= 239 - 66.4\, e^{-3.27\,t_{g1}}\left(\cosh 3.1\,t_{g1} + 1.05 \sinh 3.1\,t_{g1}\right) \\
&\quad - 172.6\, e^{-0.033\,t_{g2}}\left(\cos t_{g2} + 0.033 \sin t_{g2}\right) \\
&= v_{SR}.
\end{aligned}$$

The result shows that a duality exists between the series reactor representation in a sourced versus source-free circuit. With reference to Tables 2.1 and 2.2, the evident relationship between the equations for the series circuit with a precharged capacitor and those for the sourced circuit with current injection also suggests the existence of a latent duality between the two circuits. The duality is described in Table 7.1.

Given the above calculation and its result and the discussion of duality, we can reverse engineer the result to a general equation for the circuit breaker TRV where series reactors are applied. For the case of an overdamped source TRV, the equation is

Table 7.1 Duality between source- and load-side series reactor representation.

Source side	Load side
Parallel RLC with current injection	Series RLC with precharged capacitor
Oscillation starting point = 0	Oscillation starting point = 1 pu
Axis of oscillation = 1 pu	Axis of oscillation = 0

$$v = (V_1 + V_2) - V_1 e^{-d_p t_{g1}} \left(\cosh \sqrt{d_p^2 - 1} \, t_{g1} + \frac{d_p}{\sqrt{d_p^2 - 1}} \sinh \sqrt{d_p^2 - 1} \, t_{g1} \right)$$

$$- V_2 e^{-d_s t_{g2}} \left(\cos \sqrt{1 - d_s^2} \, t_{g2} + \frac{d_s}{\sqrt{1 - d_s^2}} \sin \sqrt{1 - d_s^2} \, t_{g2} \right),$$

where V_1 is the voltage drop across the source impedance before current interruption, V_2 is the voltage drop across the series reactor before current interruption, t_{g1} is the generic time for the source circuit TRV representation, t_{g2} is the generic time for the series reactor TRV representation, and d_p and d_s are degrees of damping as discussed in Chapter 2 and $d_p = d_s$ for the series reactor circuit.

If the source TRV is underdamped, the second term in the equation is replaced with

$$V_1 e^{-d_p t_{g1}} \left(\cos \sqrt{1 - d_p^2} \, t_{g1} + \frac{d_p}{\sqrt{1 - d_p^2}} \sin \sqrt{1 - d_p^2} \, t_{g1} \right).$$

These considerations show that the effect of series reactor can be treated by applying superposition theory to the source and series reactor individually or by separating them and again treating them individually. This shows that the decoupling and conversion Boehne–Hammarlund method, quite apart from the fact that it does not consider damping, has no obvious validity.

A practical perspective on series reactors applied to limit fault current is to consider two circuit breakers connected in series with a series reactor in between. Such a case is where a bus circuit breaker on a stepdown transformer secondary connects through a series reactor to a feeder circuit breaker. In the event of a feeder fault cleared by either circuit breaker, each is exposed to the same TRV.

7.2.2 De-Energization

7.2.2.1 General Considerations

Capacitive current interruption has its own unique features, and the main subjects for consideration are recovery voltages and re-striking currents and overvoltage:

- *Recovery voltages*: The TRV across the circuit breaker is the difference between the source-side AC voltage and the load-side trapped DC voltage. Being the slowest of all TRVs, it is best described as a recovery voltage rather than as a TRV.
- *Re-striking*: Re-striking is equivalent to capacitor bank insertion, actually re-insertion, and results in inrush currents and possible escalating trapped DC voltages on the banks.

7.2.2.2 Recovery Voltages

Two cases are of interest, shunt capacitor banks with earthed neutrals and those with unearthed neutrals.

Capacitive Load Switching

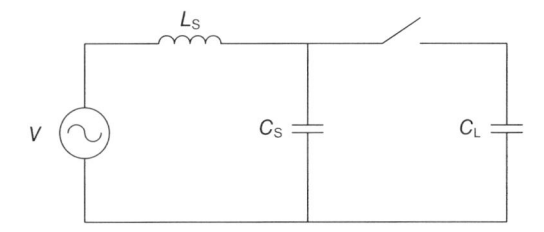

Figure 7.12 Single-phase circuit for single capacitor bank with earthed neutral.

The recovery voltage for the first case is as shown in Figure 7.1 for 50 Hz. Each phase is stressed by the same 2 pu voltage peak one-half cycle after current interruption. Clearly, a 60 Hz power frequency is the more onerous and can be taken to also cover 50 Hz but not vice versa. Figure 7.12 shows the single-phase circuit for this case, L_S being the source inductance and C_S the source-side capacitance. At the instant of current interruption, the voltage at C_S drops at a rate determined by L_S and C_S. This drop appears as an initial component of the recovery voltage and is referred to as the voltage jump. The magnitude of the jump is approximately equal to the ratio of the capacitor bank Mvar rating to the short-circuit MVA at the bank location.

For the case of shunt capacitor banks with unearthed neutrals, first pole interruption occurs in one pole, and the currents in the other two phases shift to become equal in magnitude and opposite in polarity and are interrupted 90° after the first pole. This is shown in Figure 7.13 on a per-unit voltage basis. The sequence is as follows:

- a-phase current is interrupted first leaving a 1 pu trapped charge voltage on the a-phase capacitor bank.
- The voltage at the neutral point starts to rise, reaching a trapped charge voltage value of 0.5 pu 90° later when the b-phase and c-phase currents are interrupted.

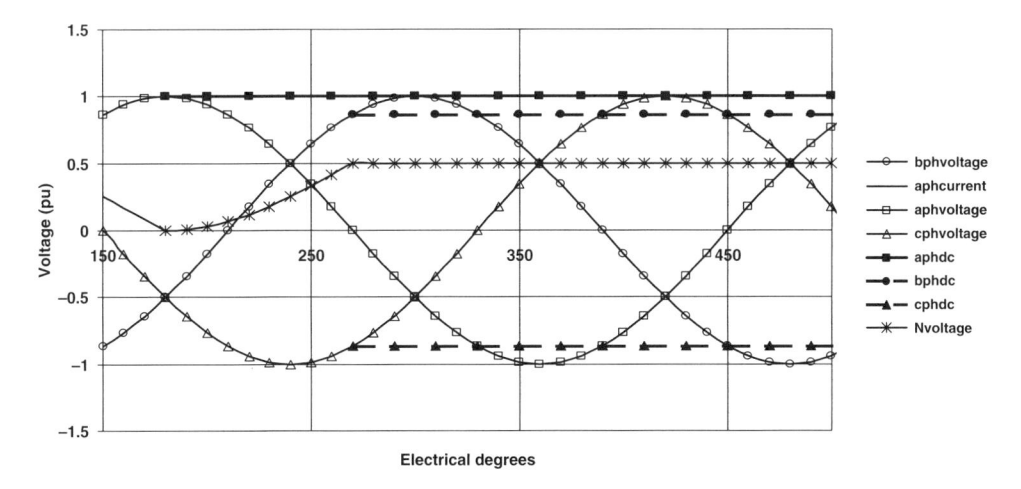

Figure 7.13 Current interruption for shunt capacitor bank with unearthed neutral.

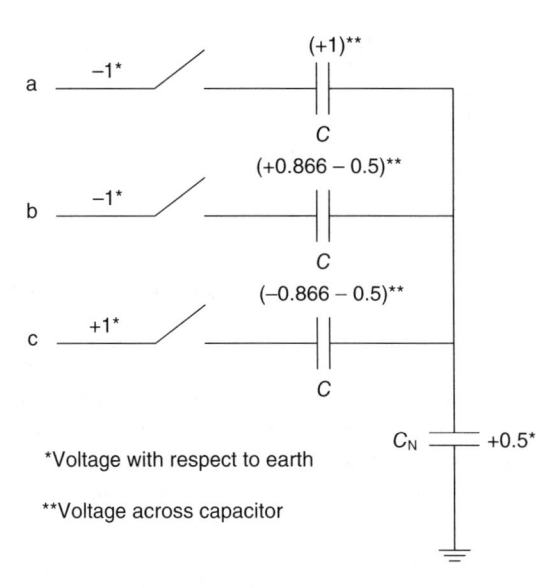

Figure 7.14 Applied and trapped charge voltage distributions for recovery voltage peak value determination.

- The trapped charge voltages on the b-phase and c-phase capacitor banks will be the difference between the applied voltage at the instant of current interruption, +0.866 and −0.866 pu, respectively, and the trapped charge voltage on the neutral point.

Figure 7.14 shows the distribution of the applied voltages and the trapped charge voltages after current interruption in all phases corresponding to the peak values of the recovery voltages in each phase. Referencing the voltages on either side of the circuit breaker to the same point (earth), the recovery voltage peak values are 2.5 pu for a-phase and 1.866 pu for b- and c-phases, but displaced in time as shown in Figure 7.15.

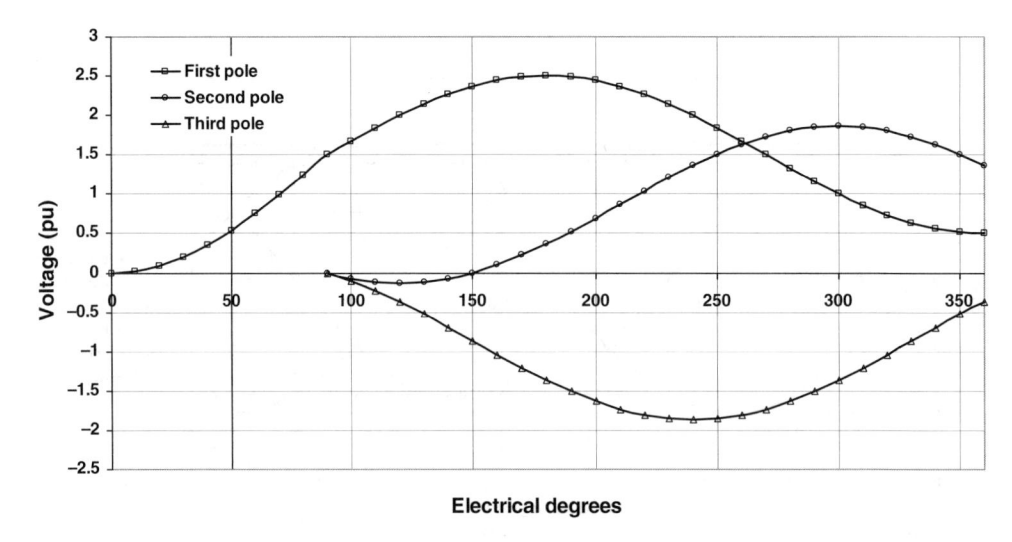

Figure 7.15 Recovery voltages for shunt capacitor bank switching on a-, b- and c-phases.

Capacitive Load Switching

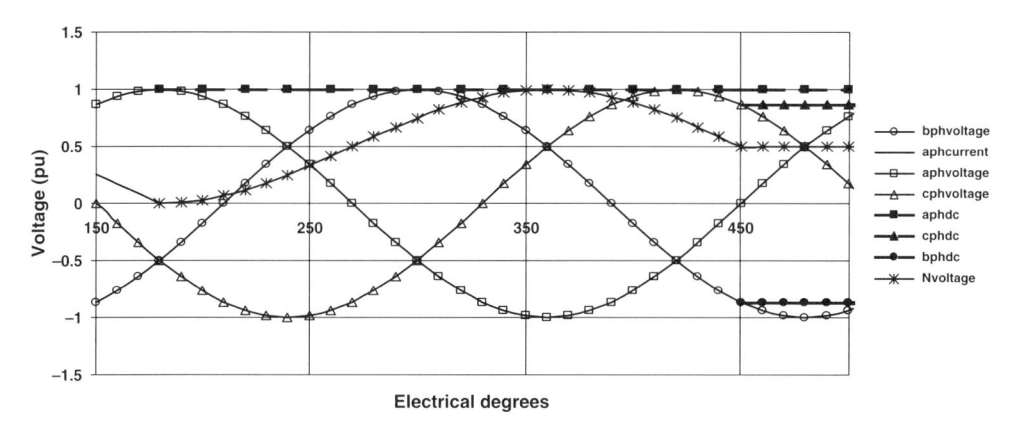

Figure 7.16 Current interruption for shunt capacitor bank with unearthed neutral with delayed interruption in the second and third poles.

An extension of this case is where the second and third phases do not interrupt the current at 90° but are delayed and conduct current for a further 180° (Figure 7.16). The excursion of the neutral voltage now continues to 1 pu and then back down to a final 0.5 pu trapped voltage.

Polarities of the voltages on the other two phases reverse compared with 90° current interruption. The distribution of the applied and trapped charge voltages in each phase corresponding to the peak values of the recovery voltage is shown in Figure 7.17. Figure 7.18 shows that the recovery voltage peaks on a-phase are now 3 pu. The recovery voltage peaks on b- and c-phases remain at 1.866 pu but now delayed by 180° in time.

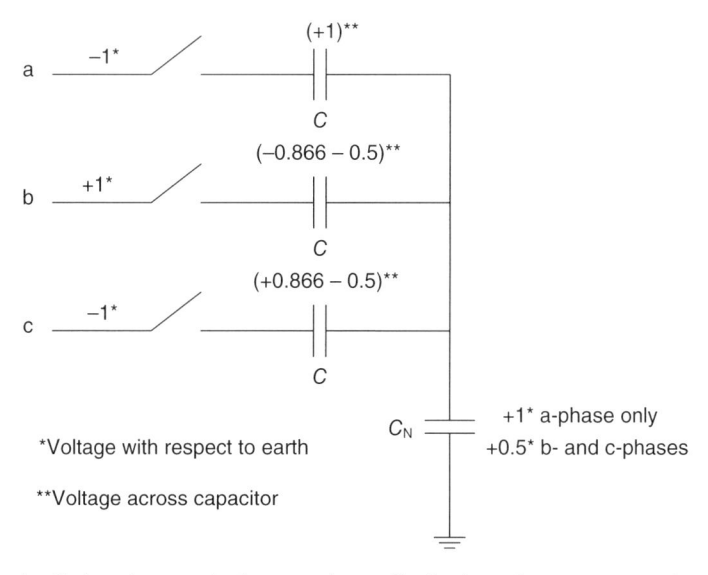

Figure 7.17 Applied and trapped charge voltage distributions for recovery voltage peak value determination with delayed interruption in the second and third poles.

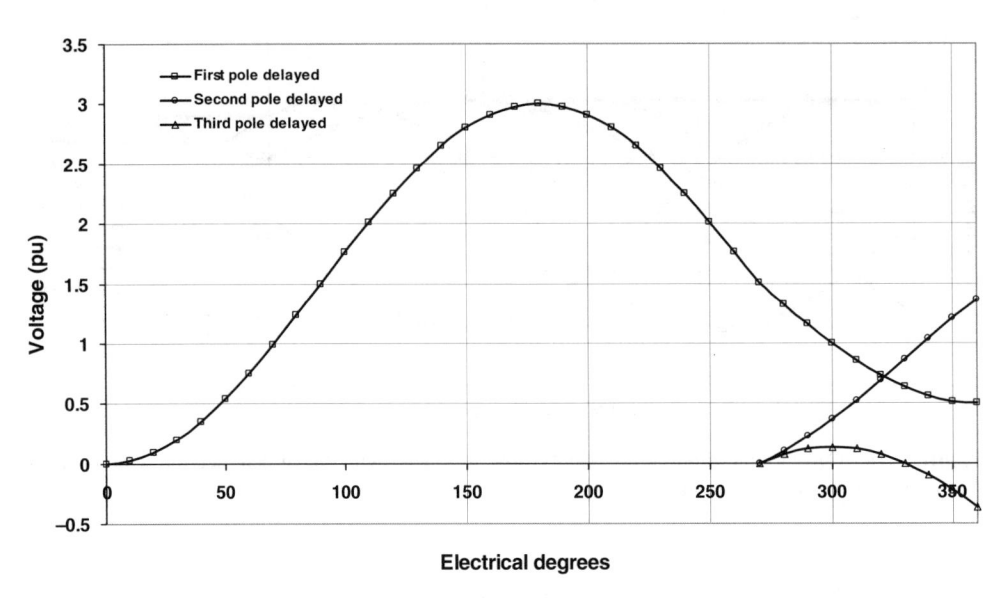

Figure 7.18 Recovery voltage for a-phase with delayed interruption in the second and third poles.

7.2.2.3 Re-ignitions and Re-strikes

Re-ignitions and re-strikes are defined by their time of occurrence. With reference to Figure 7.19, a re-ignition is defined as a voltage breakdown between the opening circuit breaker contacts within one-quarter cycle of current interruption. In turn, a re-strike is defined

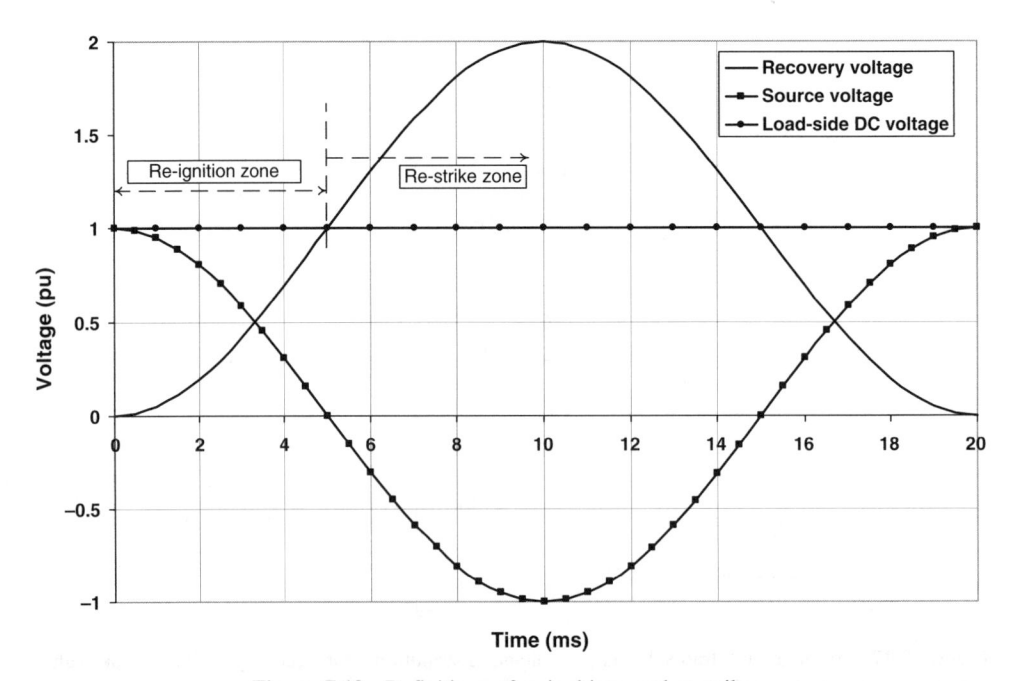

Figure 7.19 Definitions of re-ignitions and re-strikes.

Capacitive Load Switching

as a voltage breakdown one-quarter cycle or later between the circuit breaker contacts. Either event will not necessarily lead to the circuit breaker conducting another half-cycle of power frequency current because the associated re-ignition or re-strike transient currents may be interrupted by the circuit breaker.

The issues to be addressed in connection with re-ignitions and re-strikes are the overvoltages generated by the events and the consequences for the circuit breaker. For the first quarter cycle after current interruption—the re-ignition zone—the AC source-side voltage and the load-side trapped charge voltage have the same polarity. In the event of a re-ignition, the load-side voltage recovers to the source side and, the circuit being an underdamped series RLC circuit, overshoots to a value less than twice the original difference voltage across the circuit breaker. The overvoltage peak with respect to earth is not significant, and re-ignitions are generally not a major application consideration.

Re-striking is a major application event. From Figure 7.19, the voltage difference across the circuit breaker is now 1 pu or greater, and overvoltage peak values become significant. Worse still, if the re-striking current is interrupted by the circuit breaker after one loop (half-cycle), escalation of the load-side voltage can occur. This escalation is illustrated in Figure 7.20, assuming no damping in the circuit. The initial trapped charge voltage is 1 pu positive, and a re-strike at the next source-side voltage negative peak will result in a -3 pu voltage on the capacitor bank. This assumes interruption of the re-striking current as the re-striking voltage oscillation reaches its peak value. The result is evident if we remember the logic learned in Chapter 2: the starting point of the oscillation is the 1 pu DC voltage on the capacitor bank, the axis of oscillation (aiming point) is the AC source voltage peak and the peak value relative to the starting point is twice the original difference voltage across the circuit breaker (amplitude factor equal to 2 for no damping). At the next source-side positive peak, the recovery voltage

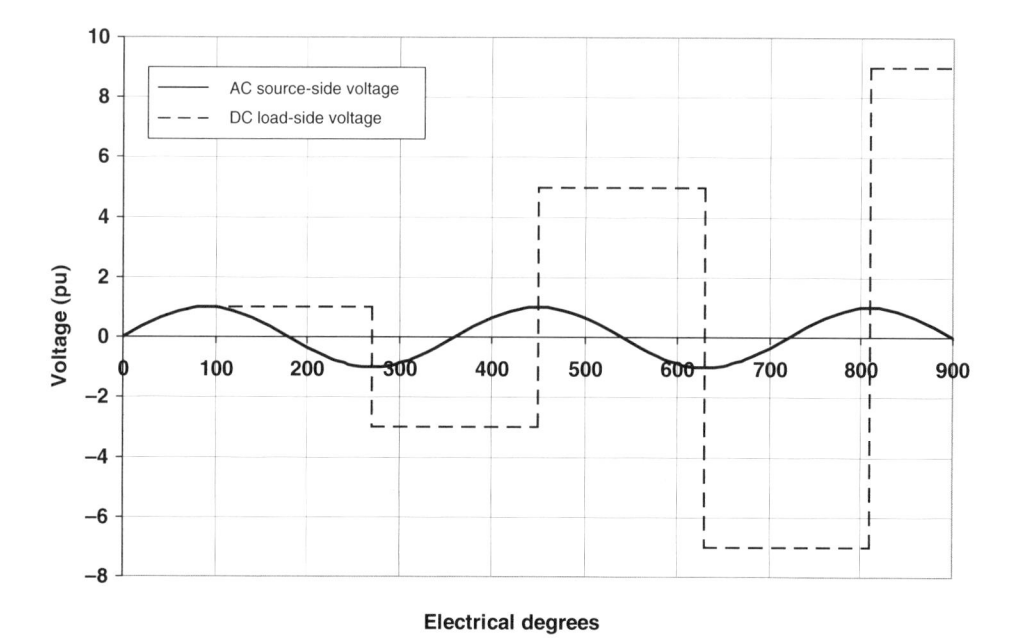

Figure 7.20 Load-side voltage escalation caused by multiple re-striking.

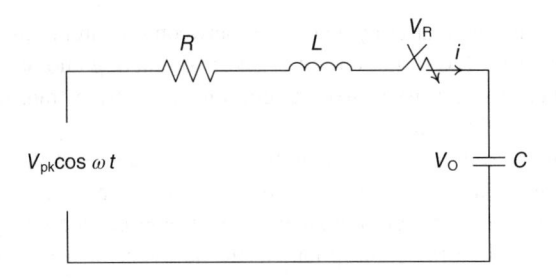

Figure 7.21 Circuit for re-striking voltage calculation.

peak is now 4 pu, and a re-strike will leave a positive 5 pu DC voltage on the capacitor bank. Clearly, this is an undesirable situation, and circuit breakers are designed to have a low or very low probability of re-striking as defined by the relevant type testing requirements in IEC 62271-100.

Re-striking can, of course, also be treated analytically by considering the circuit shown in Figure 7.21. V_R is the voltage at which the re-strike occurs and is the difference between the source voltage and the voltage on the capacitor. From the perspective of the re-striking current, this is equivalent to injecting a voltage step V_R into the RLC circuit of Figure 7.2 with a precharge V_0 on the capacitor. From Eq. (7.4), the current i is given by

$$i = \frac{V_R}{Z} e^{-\alpha t} \sin \omega_0 t,$$

where $\omega_0 = 1/\sqrt{LC}$ to distinguish it from ω, which now is the power angular frequency.

Assuming an initial positive voltage V_0 on the capacitor, the voltage v_c on the capacitor is given by

$$\begin{aligned} v_c &= V_0 + \frac{1}{C}\int i \cdot dt \\ &= V_0 + \frac{V_R}{ZC}\int e^{-\alpha t} \sin \omega_0 t \cdot dt. \end{aligned} \quad (7.11)$$

Applying the method of integration by parts, the solution for Eq. (7.11) is

$$v_c = V_0 - \frac{V_R}{ZC}\left[\left(\frac{e^{-\alpha t}}{\alpha^2 + \omega_0^2}\right)(\alpha \sin \omega_0 t + \omega_0 \cos \omega_0 t)\right] + K, \quad (7.12)$$

where K is a constant dependent on the initial conditions. Assuming zero damping, Eq. (7.12) reduces to

$$\begin{aligned} v_c &= V_0 - \frac{V_R}{ZC\omega_0} \cos \omega_0 t + K \\ &= V_0 - V_R \cos \omega_0 t + K. \end{aligned} \quad (7.13)$$

At $t = 0$, $v_c = V_0$ giving $K = V_R$ and v_c as

$$v_c = V_0 + V_R(1 - \cos \omega_0 t).$$

Capacitive Load Switching

The term in the brackets has a maximum value of 2, and

$$v_{c\,max} = V_0 + 2V_R. \tag{7.14}$$

In using Eq. (7.14), it is important to keep track of polarity. With reference to Figure 7.20, $V_0 = +V_{pk}$ at current interruption and (remembering that V_R is the source minus the load voltages) 180° later $V_R = -2V_{pk}$ and

$$v_{c\,max} = V_{pk} + 2(-2V_{pk})$$
$$= -3V_{pk}.$$

Taking another 180° step,

$$v_{c\,max} = -3V_{pk} + 2(V_{pk} - (-3V_{pk}))$$
$$= -3V_{pk} + 8V_{pk}$$
$$= 5V_{pk}.$$

For the general case, K equals the bracketed term in Eq. (7.12), giving

$$v_c = V_0 + V_R \left[1 - \left(\frac{e^{-\alpha t}}{1 + (\alpha/\omega_0)^2}\right)\left(\cos \omega_0 t + \frac{\alpha}{\omega_0}\sin \omega_0 t\right)\right].$$

The re-striking current is given by

$$i = \frac{V_R}{Z} e^{-\alpha t} \sin \omega_0 t.$$

Readers are encouraged to prove this by calculating the rate of change of the charge on the capacitor.

7.2.3 Outrush

Outrush is the discharge of a shunt capacitor bank through a circuit breaker (or circuit breakers) in the process of clearing a fault. The situation is as illustrated in Figure 7.22. The discharge current is the quantity of interest and is given by Eq. (7.4).

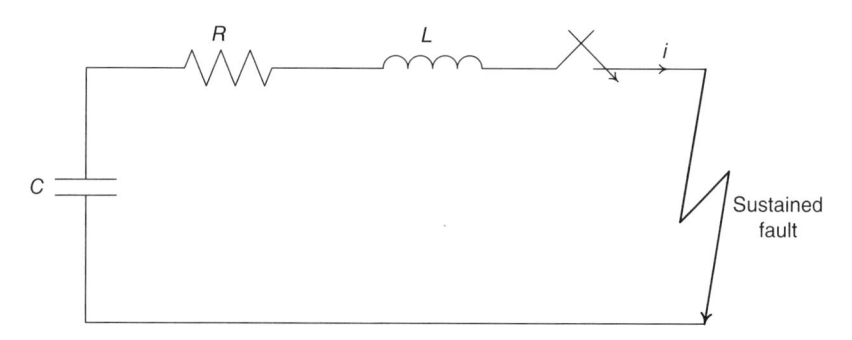

Figure 7.22 Circuit for shunt capacitor bank outrush current calculation.

A possible sequence of events is as follows assuming auto-reclosing of the circuit breaker:

- On initiation of the fault, the capacitor bank discharges through the circuit breaker. At this point, the circuit breaker is still in the closed position.
- The circuit breaker is tripped and will attempt to interrupt the fault current at the first or at a later zero crossing. A failed attempt to interrupt means the TRV builds up before a re-ignition, charging the bank and then discharging the bank when the re-ignition occurs. The fault is cleared at the next current zero, and the bank resumes normal service.
- The circuit breaker recloses after a certain time interval. If the fault is sustained, the circuit breaker will prestrike again, discharging the bank. The circuit breaker will close fully and then trip, repeating the interrupting event described above.

The issue with outrush is that the discharge current superimposes on the fault current and may force a premature current zero and current interruption. This can be repeated several times, leading to voltage escalation and consequential failure of the circuit breaker. Such failures are very rare, being most probable on circuit breakers prone to dielectric re-ignitions. Older bulk-oil and minimum-oil circuit breakers fall into this category. Vacuum circuit breakers also have this characteristic, but few as yet are applied at transmission level system voltages.

7.3 Transmission Lines

Unloaded transmission lines represent a capacitive load but of a more complex nature than shunt capacitor banks. The line capacitance is distributed, rather than lumped, and involves both line-to-earth and line-to-line capacitance. Lines are characterized by their positive- (C_1) and negative-sequence (C_0) capacitances connected as shown in Figure 7.23. $C_1 = C_0$ is the base case having only capacitance to ground as for a shunt capacitor bank with an earthed neutral.

Transmission lines have high surge impedance in the order of hundreds of ohms; therefore, inrush currents and back-to-back line switching are of little or no significance. The major issue with the closing (energization) or reclosing on transmission is the switching surges that are generated. These surges stress the line insulation and are one of the main design parameters at system voltages of 300 kV and above. Usually some form of mitigation is applied to limit the magnitude of the switching surges.

The recover voltage will, of course, be a $(1 - \cos)$ function but with a peak value (k_{rv} in pu) dependent on the ratio C_1/C_0. The quantity to be derived is the first-pole-to-clear factor (k_{pp} in pu) referenced to the base case of $C_1 = C_0$ and a load impedance of $X_1 = 1/\omega C_1$ (Figure 7.24). The k_{pp} factor also has to be applied to the supply base voltage to achieve the correct capacitance load current. From Figure 7.1, we know that $k_{rv} = 2k_{pp}$.

Given the line representative circuit of Figure 7.23, and with reference to Appendix F, the circuit could be converted (albeit with some effort) to the format of Figure F.1. However, this is not necessary because the calculation in Appendix F shows that in general first-pole-to-clear representation is by earthing the second and third phases as shown in Figure 7.25. From the right circuit in Figure 7.24, we derive the load impedance X_L through circuit reduction, giving

$$X_L = \frac{1.5 X_0 X}{1.5 X + X_0},$$

where $X = 1/\omega(C_1 - C_0)$ and $X_0 = 1/\omega C_0$.

Capacitive Load Switching

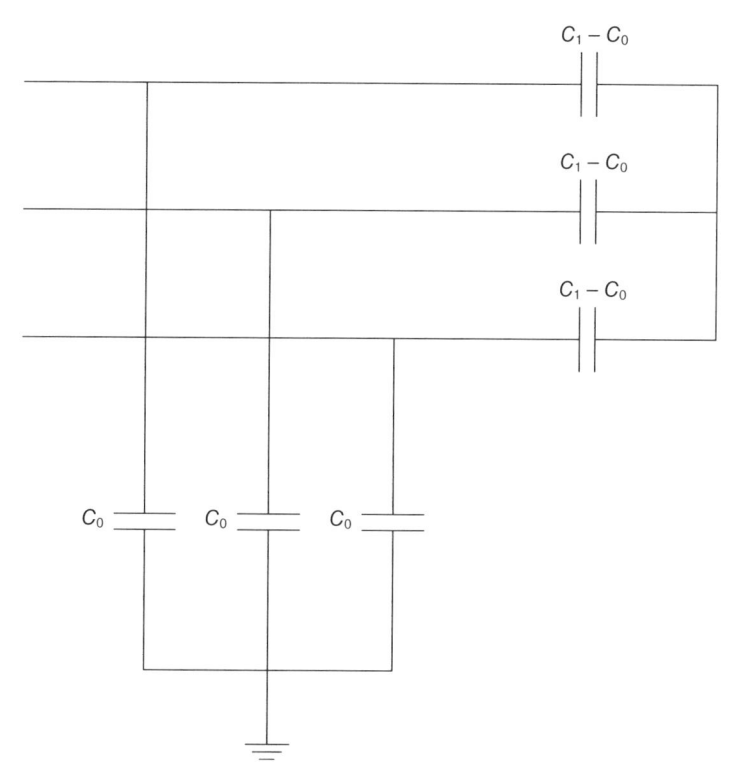

Figure 7.23 Representative circuit for unloaded transmission lines.

By reference to Figure 7.24,

$$X_L = k_{pp} X_1$$

or

$$k_{pp} = X_L/X_1 = \frac{1.5(C_1/C_0)}{0.5 + (C_1/C_0)}$$

Figure 7.24 Single-phase circuit representation for line switching.

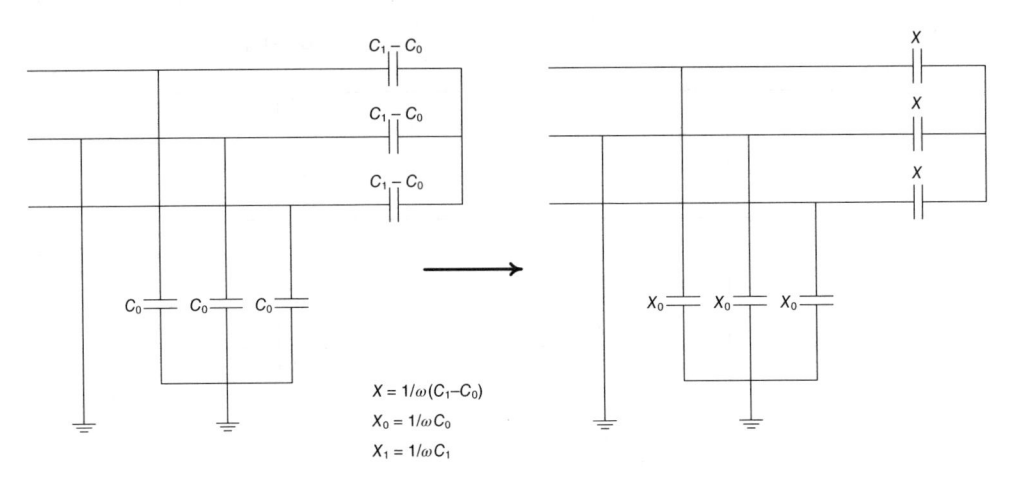

Figure 7.25 Circuit for transmission line first-pole-to-clear calculation.

after conversion back to the capacitance value. Finally, then

$$k_{rv} = \frac{3(C_1/C_0)}{0.5 + (C_1/C_0)}. \tag{7.15}$$

Equation (7.15) is plotted in Figure 7.26 converging toward 3 pu. As expected, the base case ($C_1/C_0 = 1$) gives $k_{rv} = 2$. Typically, transmission lines have a C_1/C_0 ratio equal to 2, and k_{rv} is

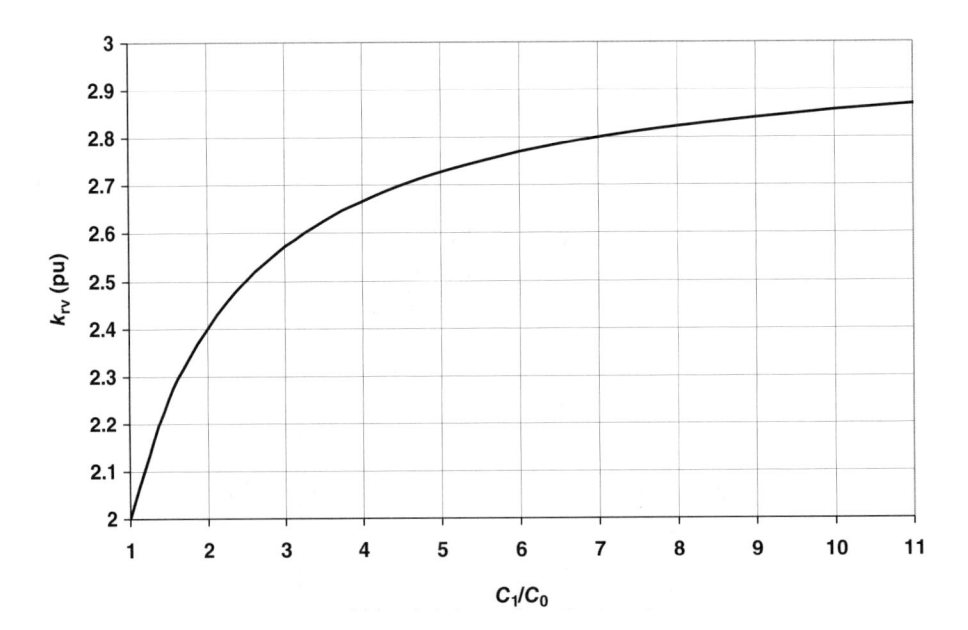

Figure 7.26 Transmission line recovery voltage peak versus C_1/C_0 ratio.

Capacitive Load Switching

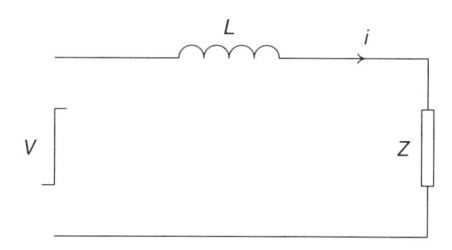

Figure 7.27 Circuit for unloaded cable inrush current calculation.

2.4 pu. Other values of k_{rv} are also used in type testing to reflect particular circumstances as discussed in Section 7.5.

7.4 Cables

The circuit for single cable switching is shown in Figure 7.27. The cable is represented by its surge impedance and, as we learned in Chapter 4, a surge impedance is resistive. Given that there is no capacitance in the circuit, the current can only be overdamped. Cables have low surge impedances in the order of tens of ohms, and inrush current magnitudes can be of significance. Note that the calculation that follows is also applicable to transmission lines, but as also noted earlier, line surge impedances are too high to permit inrush currents of any interest.

From Figure 7.27, we can write

$$L\frac{di}{dt} + Z_i = V,$$

where V is the injected step voltage.

Differentiating across gives

$$\frac{d^2i}{dt^2} + \frac{Z}{L}i = 0. \tag{7.16}$$

Equation (7.16) is in our *abc* format with $c = 0$ and $\alpha = \beta = Z/2L$.

From Eq. (2.4) and ignoring damping,

$$i = \frac{V}{L\alpha}e^{-\alpha t}\sinh \alpha t$$
$$= \frac{2V}{Z}e^{-\alpha t}\left[\frac{e^{\alpha t} - e^{-\alpha t}}{2}\right]$$
$$= \frac{V}{Z}\left(1 - e^{-2\alpha t}\right),$$

giving

$$i = \frac{V}{Z}\left(1 - e^{-Zt/L}\right). \tag{7.17}$$

The peak current magnitude (I_{pk}) is V/Z, noting that if the cable is precharged, then V is the difference between the applied voltage and the trapped voltage on the cable (which may have opposite polarities). The limiting peak current value is 20 kA, and we need to calculate an equivalent frequency for comparison with the limiting frequency of 4250 Hz.

From Eq. (7.17),

$$\frac{di}{dt} = \frac{V}{L} e^{-Zt/L}$$
$$= \frac{V}{L} \quad \text{at } t = 0. \tag{7.18}$$

Equation (7.18) gives a rate of change of current, which we can relate to the limiting peak inrush current:

$$\frac{di_i}{dt} = 2\pi f_{ie} I_{ipk}, \tag{7.19}$$

where i_i is the instantaneous inrush current, I_{ipk} is the inrush current peak and f_{ie} is its equivalent frequency.

Equating Eqs. (7.18) and (7.19) gives

$$f_{ie} = \frac{V}{2\pi L I_{ipk}}, \tag{7.20}$$

and Eqs. (7.17) and (7.20) can now be used to determine if a particular application meets the standard requirements or if some form of mitigation is required.

The circuit for back-to-back cable switching is shown in Figure 7.28, and readers can deduce (or calculate) that the associated peak current and equivalent frequency, respectively, are given by

$$I_{pk} = \frac{V_1 - V_2}{Z_1 + Z_2}$$

and

$$f_{ie} = \frac{V_1 - V_2}{2\pi L I_{ipk}}.$$

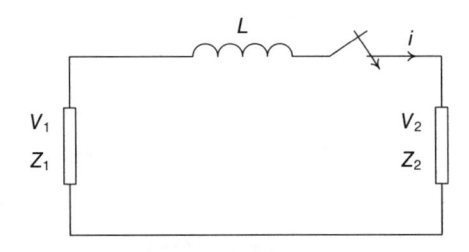

Figure 7.28 Circuit for back-to-back cable switching.

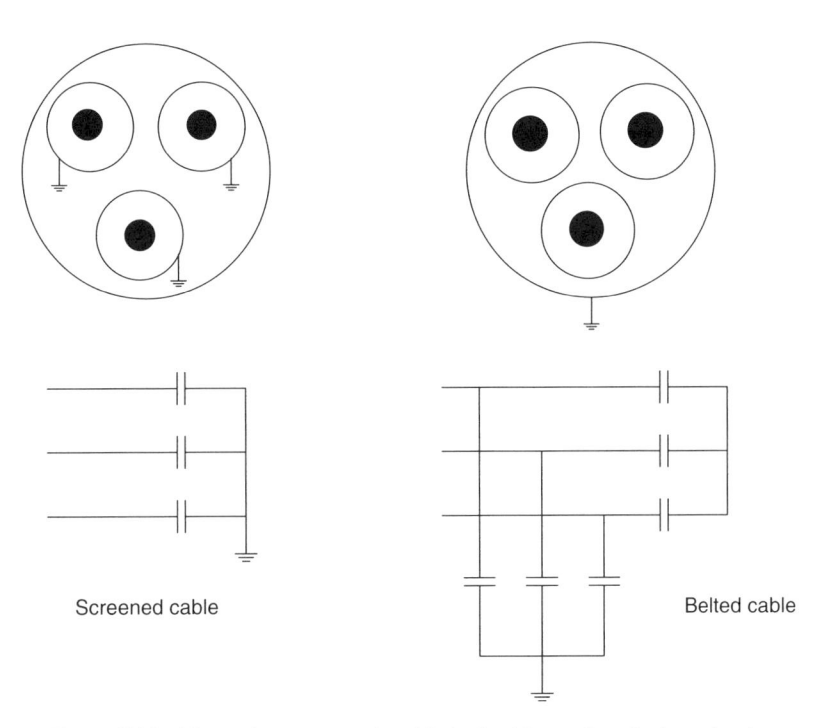

Figure 7.29 Three-phase screened and belted cables and equivalent circuits.

Three-phase cables can be either screened or belted as illustrated in Figure 7.29 together with their equivalent circuits. Screened cables can be treated in the same manner as for shunt capacitor banks with earthed neutrals and belted cables as for unloaded transmission lines and banks with unearthed neutrals as applicable (such banks have an equivalent C_1/C_0 ratio of 2.5 based on a 2.5 pu recovery voltage). For the latter case, because of the low surge impedance of cables, inrush current magnitudes may be high enough to require some form of mitigation.

7.5 Summary

The main issues with making and breaking capacitive currents are inrush currents and the circuit breaker re-striking performance, respectively.

Both the IEC and IEEE circuit breaker standards limit inrush current to 20 kA peak and a maximum allowable frequency of 4250 Hz. Circuit breakers are type tested on this basis, and the user should implement some form of mitigation if required in the particular application. The mitigation choices and their pros and cons are detailed in Table 7.2.

Circuit breaker re-striking performance is statistical, and the standards now have two capacitive current switching classes of C1 and C2. A class C1 circuit breaker is defined as one "with a low probability of re-strike during capacitive current breaking as demonstrated by a specific type test." A class C2 circuit breaker has a similar definition except that it has a "very low probability of re-strike" in a specific type test. Obviously, the C2 type test in terms of the number of test shots and the acceptance will be more onerous than the C1 case. The test requirements are very detailed (and evolving), and readers are referred to the latest edition of the standards.

Table 7.2 Inrush current mitigation measures.

Mitigation measure	Pros	Cons
Fixed series reactors	Passive devices; limit inrush, re-ignition, re-striking and outrush currents	Losses; require space
Pre-insertion inductors or resistors	Only in circuit when required; no losses and no extra space required	Active devices; may require maintenance; do not limit re-ignition, re-striking or outrush currents
Controllers (controlled closing devices)	Installed in control building; many proven devices in the market	Active devices; may require maintenance; do not limit re-ignition, re-striking or outrush currents; require single pole operators

Table 7.3 Voltage test factors for single-phase capacitive current switching tests.

Voltage test factor	Case
1.0	Capacitor banks with earthed neutrals and screened cables in effectively earthed systems with no significant coupling with adjacent phases.
1.2	Belted cable and unloaded transmission line switching on effectively earthed systems with significant coupling with adjacent phases for rated voltages of 52 kV and above.
1.4	1. Capacitor banks with unearthed neutrals. 2. Belted cable and unloaded line switching on effectively earthed systems for rated voltages less than 52 kV. 3. Breaking capacitive current in sound phase or phases during single-phase or two-phase fault to earth in effectively earthed systems.
1.7	Breaking capacitive current in sound phase or phases during single-phase or two-phase faults to earth in non-effectively earthed systems.

For single-phase testing, various voltage test factors are applied in order to correctly represent the first-pole-to-clear recovery voltage. The factors are listed in Table 7.3.

The use of disconnectors to interrupt small capacitive currents is a special case and is discussed in Appendix G. The disconnectors can be the air-break type used in air-insulated substations and the SF_6 gas type used in gas-insulated substations (GIS).

Bibliography

1. IEC62271-100 (2012) *High-Voltage Switchgear and Controlgear – Part 100: Alternating Current Circuit-Breakers*, International Electrotechnical Commission.
2. IEC62271-302 (2012) *High-Voltage Switchgear and Controlgear – Part 302: Alternating Current Circuit-Breakers with Intentionally Non-Simultaneous Pole Operation*, International Electrotechnical Commission.

8

Circuit Breaker Type Testing

8.1 Introduction

To this point we have considered current interruption transients without regard to the interaction with circuit breakers and their capability to perform the various switching duties. This capability has to be proven by type testing, which covers both the mandatory type tests to be performed on all circuit breakers and the type tests related to non-standardized applications such as shunt reactor switching.

The mandatory type tests according to IEC 62271-100 relate to short-circuit current interruption. This involves a number of component ratings to be proven as listed and explained in Table 8.1. As discussed in Chapter 4, the short-circuit current interrupting capability has to be demonstrated over a range from the 100% short-circuit current rating down to the 10% value with varying transient recovery voltage (TRV) values.

The application-related current interruption type tests according to IEC 62271-100 are listed and explained in Table 8.2. The listing includes both short-circuit current interruption tests (e.g. the short-line fault and out-of-phase switching) and load current switching tests.

For the mandatory and other high-current breaking tests, the intent is to demonstrate the circuit breaker capability to interrupt fault currents and recover against the prescribed TRVs, all within arcing window and interrupting time limits (refer to Figure 1.1). This is discussed further in the next section.

For the inductive and capacitive load current switching tests, the capability of the circuit breaker to interrupt the low currents is not in question. In fact, we have seen that, for shunt reactor switching, the current is forced to a premature zero, which is also a possibility with capacitive currents. Breaking reactive load currents tends to be interactive with the circuit, and testing demonstrates the performance of the circuit breaker in such an environment. This is discussed further in Sections 8.3 and 8.4.

8.2 Circuit Breaker Interrupting Time

All circuit breakers have a rated interrupting time, which is the sum of the mechanical opening time and the maximum arcing time. The last-pole-to-clear is the pole exposed to the maximum arcing time (Figure 1.1).

Current Interruption Transients Calculation, First Edition. David F. Peelo.
© 2014 John Wiley & Sons, Ltd. Published 2014 by John Wiley & Sons, Ltd.

Table 8.1 Short-circuit making and breaking mandatory type testing.

Rating component	Definition and purpose
Rated short-time withstand current (I_k)	The rms value of the current that the circuit breaker can carry in the closed position during a specific short time under prescribed conditions of use and behaviour. The rated short-time current is equal to the rated short-circuit breaking current and demonstrates the capability of the circuit breaker to carry through faults.
Rated peak withstand current (I_p)	The peak current associated with the first major loop of the rated short-circuit current that the circuit breaker can carry in the closed position under prescribed conditions of use and behaviour. The rated peak withstand current is equal to the rated short-circuit making current and demonstrates the capability of the circuit breaker to carry the peak asymmetrical currents discussed in Section 4.7.
Rated duration of short-circuit (t_k)	The time interval for which the circuit breaker can carry, in the closed position, a current equal to its short-time withstand current. The standard value of rated short-circuit duration is 1 s.
Rated short-circuit breaking current (I_{sc})	The rated short-circuit breaking current is the highest short-circuit current that the circuit breaker shall be capable of breaking under the conditions of use and behaviour prescribed in IEC 62271-100. The rated short-circuit breaking current is characterized by two values: • The rms value of its AC component (symmetrical current). • The DC time constant of the short-circuit breaking current, which results in a percentage of DC component at contact separation (asymmetrical). Conditions of use and behaviour include rated supply voltages for auxiliary and control circuits and filling pressures for gas circuit breakers.
Rated short-circuit making current	The rated short-circuit making current is a peak value derived from the rms AC short-current rating value and the applicable DC time constant (refer to Section 4.7): • 50 Hz and 45 ms: $2.5 \times I_{sc}$. • 60 Hz and 45 ms: $2.6 \times I_{sc}$. All cases in which the time constant is greater than 45 ms: $2.7 \times I_{sc}$.

Taking first the case of circuit breakers applied on non-effectively earthed systems, we can recap as follows:

- Circuit breakers are short-circuit current rated on the basis of clearing a three-phase unearthed fault (Figure 3.7).
- The first-pole-to-clear recovers against the TRV and an AC (recovery) voltage of 1.5 pu.
- The second and third poles will clear 90° later against the TRV and an AC (recovery) voltage of 0.866 pu.

Table 8.2 Application-related current making and breaking type testing.

Type test	Requirement
Out-of-phase making and breaking tests	Required for circuit breakers applied on systems or parts of a system in which an out-of-phase condition is possible. Such conditions are not intentional but rather caused by a human or control error. Rated out-of-phase making and breaking current is standardized at 25% of the rated rms AC short-circuit current.
Short-line fault tests	Required for circuit breakers rated at $U_r \geq 15$ kV and $I_{sc} > 12.5$ and applied on systems with overhead lines and solidly earthed neutrals $(k_{pp} = 1)$.
Single-phase tests	Required to demonstrate circuit breaker capability to interrupt a single-phase fault in an effectively earthed system at the relevant parameters. Also, for circuit breakers having a common operating mechanism and a common opening release, to demonstrate that the mechanical operation is not adversely affected by unbalance forces for the single-phase fault current.
Double earth fault tests	Required for circuit breakers applied on non-effectively earthed neutral systems where the possibility exists for earth faults on two different phases, one of which occurs on one side of the circuit breaker and the other one on the other side. Such a fault is essentially a line-to-line fault but involving one circuit breaker pole only (refer to Section 4.8).
Critical current tests *Definition*: Value of breaking current, less than rated short-circuit breaking current, at which the arcing time is a maximum and is significantly longer than at rated short-circuit breaking current.	These tests are not actually related to a particular application but rather to circuit breaker performance. The tests are additional short-circuit tests to be performed and are triggered when the minimum arcing time in any of the test duties T10, T30 and T60 is one half-cycle or more than the minimum arcing for the adjacent test duty. For example, if the minimum arcing time at T30 exceeds that at T60 by one half-cycle, then the additional test is required. Note a similar provision exists for the short-time fault test. Tests are at L90 and L75 (refer to Chapter 5), and if the minimum arcing time at L75 exceeds that at L90 by a quarter cycle or more, then a further test at L60 is required.
Shunt reactor switching tests	Required for circuit breakers applied to switch shunt reactor. Such applications vary widely in terms of the shunt reactor ratings and neutral disposition, that is, unearthed, directly earthed or neutral reactor earthed (refer to Chapter 6).
Capacitive current switching tests	Required for circuit breakers applied for the following application cases: • Line charging current breaking. • Cable charge current breaking. • Single capacitor bank switching. • Back-to-back capacitor bank switching. Circuit breakers are classified as being either C1 with a low probability of re-striking or C2 with a very low probability of re-striking (refer to Chapter 7). Preferred current ratings for the above cases are stated in IEC 62271-100.

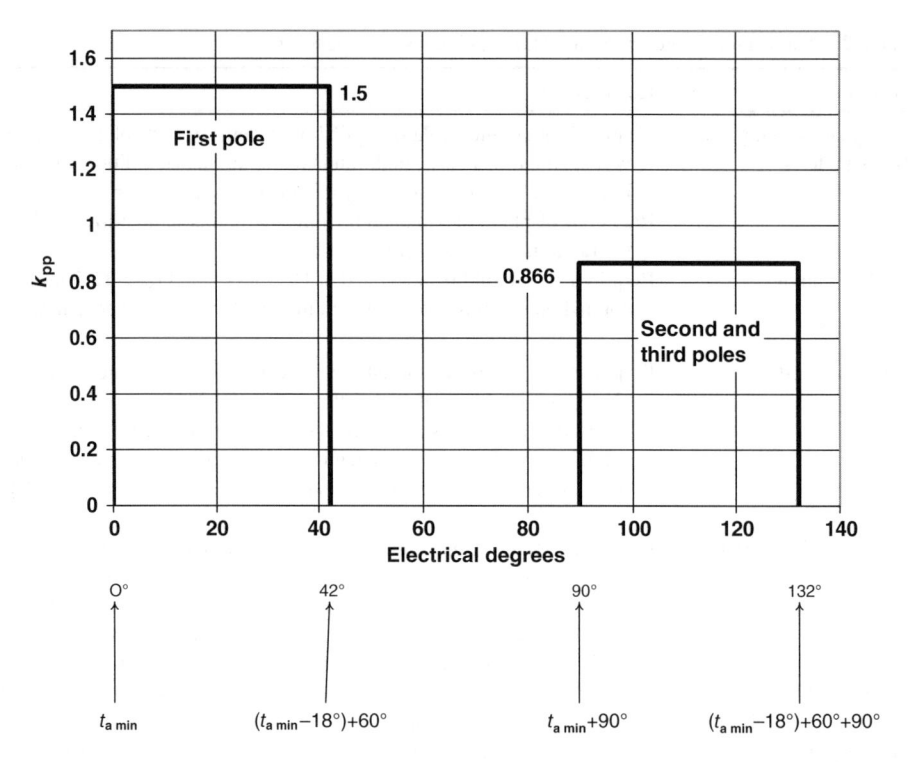

Figure 8.1 Arcing windows for circuit breakers applied on non-effectively earthed systems.

Figure 8.1 shows the arcing windows for this case. The arcing time scale starts at the minimum arcing time $t_{a\,min}$, and if contact separation on the first-pole-to-clear is such that $t_{a\,min}$ is achieved, then the second and third pole arcing times will be $t_{a\,min} + 90°$. If, as required by IEC 62271-100, the trip signal to the circuit breaker is delayed one-tenth of one half-cycle, that is, 18°, then the arcing time on the first-pole-to-clear will be

$$(t_{a\,min} - 18°) + 60° = t_{a\,min} + 42°,$$

and the arcing time on the second and third poles will be

$$(t_{a\,min} - 18°) + 60° + 90° = t_{a\,min} + 132°.$$

The maximum arcing time $t_{a\,max}$ is given by

$$t_{a\,max} = t_{a\,min} + 132°$$

and rated interrupting time by the mechanical opening time (time from applying the trip signal to contact separation) plus $(t_{a\,min} + 132°)$.

This arcing time scenario is also shown in Figure 8.2 by considering current zero crossings.

Circuit Breaker Type Testing

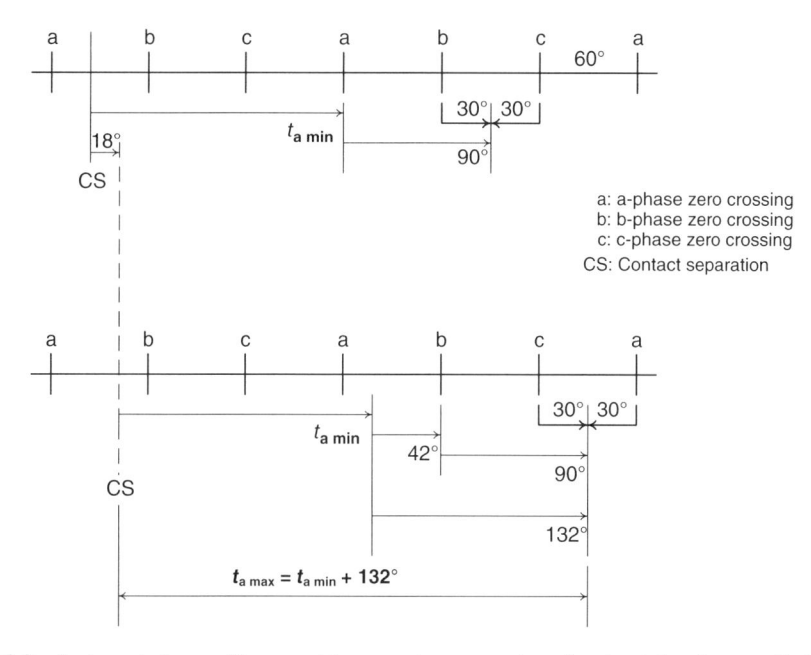

Figure 8.2 Arcing windows with respect to current zero crossings for circuit breakers applied on non-effectively earthed systems.

For the case of circuit breakers applied on effectively earthed systems, we can recap as follows:

- Circuit breakers are short-circuit current rated on the basis of clearing a three-phase-to-earth fault (Figure 3.6).
- The first-pole-to-clear recovers against the TRV and an AC (recovery) voltage of 1.3 pu.
- The second-pole-to-clear recovers against the TRV and an AC voltage of 1.27 pu.
- The third-pole-to-clear recovers against the TRV and an AC voltage of 1 pu.

Figures 8.3 and 8.4 show the arcing windows for this case. The minimum possible arcing times for the three poles are as follows (Figure 8.3):

First pole: $t_{a\,min}$.
Second pole: $t_{a\,min} + 77°$.
Third pole: $t_{a\,min} + 77° + 43° = t_{a\,min} + 120°$.

Again delaying the trip signal by 18° gives the following maximum arcing times for the three poles (Figure 8.4):

First pole: $(t_{a\,min} - 18°) + 60° = t_{a\,min} + 42°$.
Second pole: $t_{a\,min} + 42° + 77° = t_{a\,min} + 119°$.
Third pole: $t_{a\,min} + 42° + 77° + 43° = t_{a\,min} + 162°$.

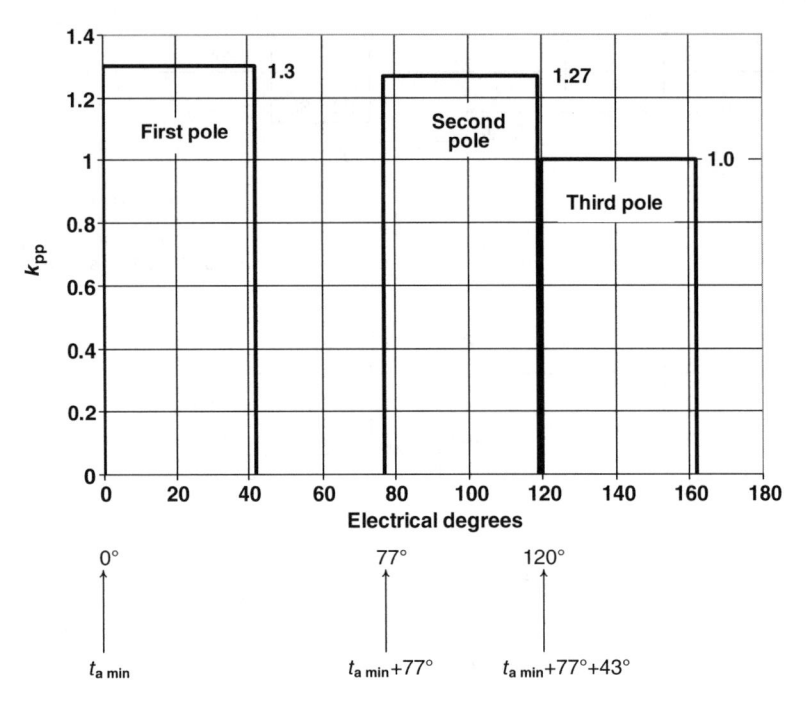

Figure 8.3 Arcing windows and minimum arcing times for circuit breakers applied on effectively earthed systems.

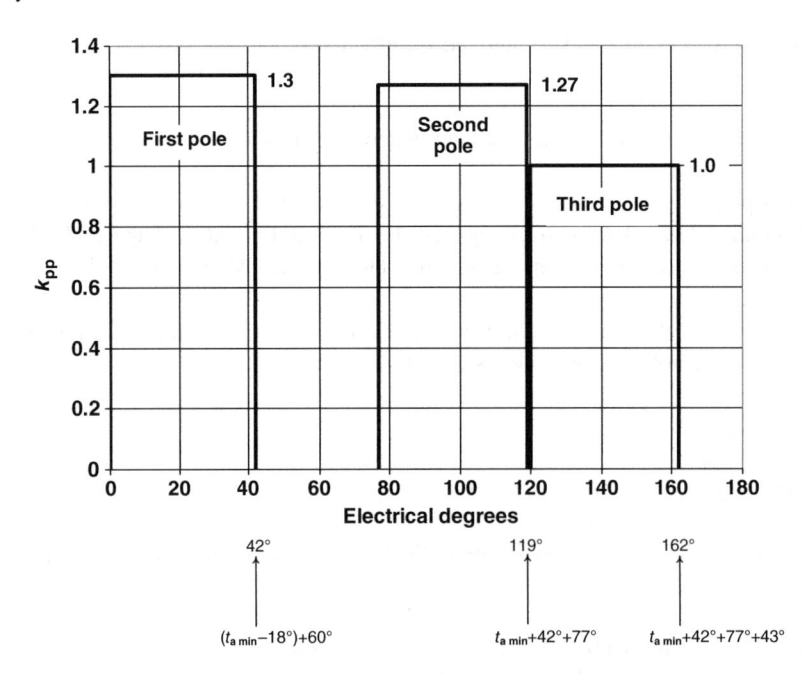

Figure 8.4 Arcing windows and maximum arcing times for circuit breakers applied on effectively earthed systems.

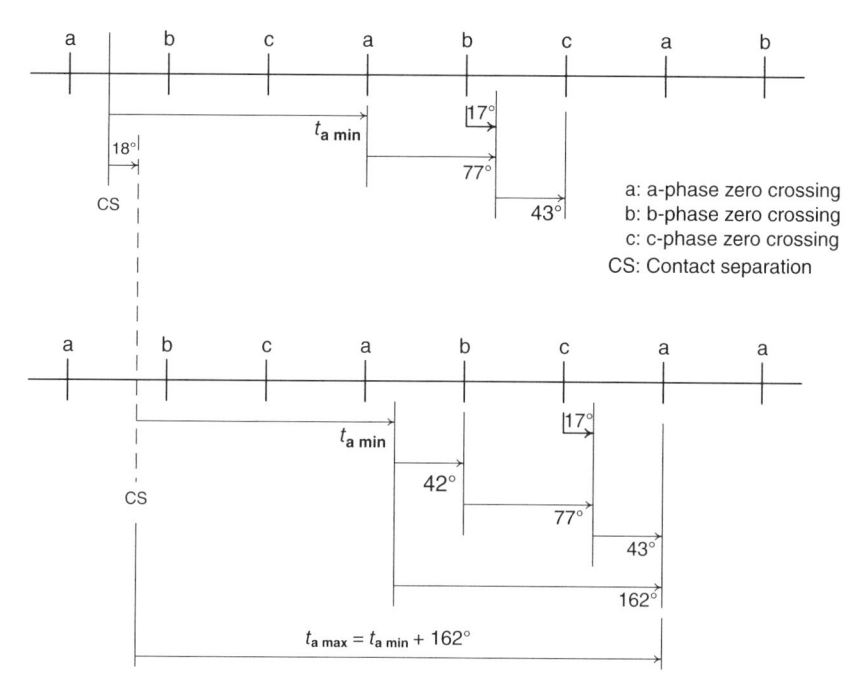

Figure 8.5 Arcing windows with respect to current zero crossings for circuit breakers applied on effectively earthed systems.

The maximum arcing time is given by

$$t_{a\ max} = t_{a\ min} + 162°$$

and rated interrupting time by the mechanical opening time plus ($t_{a\ min} + 162°$).

Figure 8.5 shows the arcing time scenario by considering current zero crossings.

For all short-circuit current interrupting tests (terminal faults, short-line faults and out-of-phase switching), the IEC standard requires that successful interruption be demonstrated at minimum and maximum arcing times and also at an intermediate arcing time.

8.3 Inherent Transient Recovery Voltages

The TRV values stated in IEC 62271-100 for the various circuit breaker voltage ratings are inherent values. This means that the test circuit is derived and set up to give the required TRV without regard for influence by the circuit breaker characteristics. Gas-type circuit breakers tend to have little or no influence on TRVs. However, for oil- and vacuum-type circuit breakers, the influence may be significant. The reason for this is that these circuit breakers exhibit post-arc current, which effectively inserts a dynamic parallel resistance into the circuit and increases the circuit damping. In contrast to gas circuit breakers, oil and vacuum circuit breakers have a low sensitivity to fast rate of rise of recovery voltages (RRRVs) as in the short-line fault test.

8.4 Inductive Load Switching

Inductive load switching covers the cases of unloaded transformer switching and shunt reactor switching. The former switching case has no significance for circuit breakers and is therefore not addressed in the circuit breaker standards. However, shunt reactor switching is a severe duty but is not standardized in terms of ratings because these are application dependent. The IEC standard covering shunt reactor switching is IEC 62271-110, which essentially describes how circuit breakers should be tested to determine their performance.

For shunt reactors applied at system voltages of 72.5 kV and above, the ratings are such that the reactor load currents are in the order of hundreds of amperes at most and less than 100 A at least. As calculated in Chapter 6, the lower the Mvar rating, and hence the lower load current, the more onerous the TRV imposed on the circuit breaker. This is shown in Figure 8.6, where it is evident that the suppression peak overvoltage increases with decreasing shunt reactor Mvar ratings. For this reason, the IEC standard requires that the test be performed at a minimum current of 100 A in general and at the actual application current if it is less than 100 A.

For the gas circuit breakers normally applied for shunt reactor switching at the above-noted system voltages, the intent of the testing is to determine performance with respect to arcing times and current chopping characteristics. The magnitude of the suppression peak voltage is dependent on the arcing time and is measured over the range of minimum to maximum arcing time. The chopping current levels are then calculated, enabling calculation of the chopping numbers, all as a function of the arcing time. The results are statistical, and analysis is required to determine the highest probable chopping numbers. The method for the statistical analysis can be found in IEC 62271-306.

Shunt reactors applied at system voltages below 72.5 kV often have similar Mvar ratings to those discussed earlier. The corresponding load currents will be an order of magnitude higher

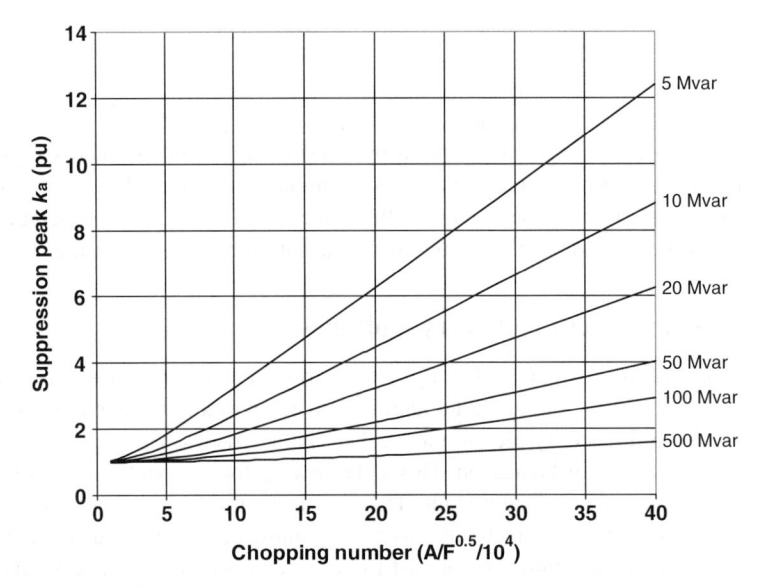

Figure 8.6 Suppression peak overvoltages versus chopping numbers for varying shunt reactor Mvar ratings.

and the intent of testing is to demonstrate current interruption. As discussed in Chapter 6, whereas the circuit characteristic of influence with respect to the TRV for gas circuit breakers in these cases is the arc voltage, that for vacuum circuit breakers is current chopping.

8.5 Capacitive Current Switching

Capacitive current switching covers the four cases noted in Table 8.2. The intent of testing is to establish the classification of the circuit breaker with respect to re-striking performance. For each application, the circuit breaker can be classified as either C1 with a low probability of re-striking or C2 with a very low probability of re-striking.

Test details are complex, and readers are referred to the latest edition of IEC 62271-100, specifically subsection 6.111.

Bibliography

1. IEC62271-100 (2012) *High-Voltage Switchgear and Controlgear – Part 100: Alternating Current Circuit-Breakers*, International Electrotechnical Commission.
2. IEC62271-110 (2012) *High-Voltage Switchgear and Controlgear – Part 110: Inductive Load Switching*, International Electrotechnical Commission.
3. IEC62271-306 (2012) *High-Voltage Switchgear and Controlgear – Part 306: Guide to IEC 62271-100, IEC 62271-1 and Other IEC Standards Related to Alternating Current Circuit-Breakers*, International Electrotechnical Commission.

Appendix A

Differential Equations

Current interruption transient oscillations in AC power systems can be treated by considering four basic RLC circuits as shown in Figure 2.1. The oscillations or other transient events that occur in these circuits caused by sudden changes are described by second-order differential equations. The equations have a common format, and rather than resorting to Laplace transforms in each case, it is possible to use a purely mathematical approach in general and then apply the general solutions to each transient case.

Two types of differential equations are applicable to the circuits shown in Figure 2.1. The first type is a second-order linear homogeneous equation with constant coefficients, which takes the form

$$a\frac{d^2x}{dt^2} + b\frac{dx}{dt} + cx = 0. \tag{A.1}$$

The solution of this equation must include two arbitrary constants and therefore in general is

$$x(t) = k_1 x_1(t) + k_2 x_2(t), \tag{A.2}$$

where k_1 and k_2 are the arbitrary constants dependent on the boundary conditions for the circuit under consideration.

The exponential function $x(t) = e^{rt}$ is a solution for Eq. (A.1), and substitution gives

$$\begin{aligned} &ar^2 e^{rt} + br\, e^{rt} + c\, e^{rt} = 0, \\ &\left(ar^2 + br + c\right)e^{rt} = 0, \\ &e^{rt} \text{ is not equal to zero and therefore} \\ &ar^2 + br + c = 0. \end{aligned} \tag{A.3}$$

Equation (A.3) is the so-called auxiliary or characteristic equation and has two roots, r_1 and r_2:

$$r_1 = \frac{-b + \sqrt{b^2 - 4ac}}{2a}, \tag{A.4}$$

Current Interruption Transients Calculation, First Edition. David F. Peelo.
© 2014 John Wiley & Sons, Ltd. Published 2014 by John Wiley & Sons, Ltd.

$$r_2 = \frac{-b - \sqrt{b^2 - 4ac}}{2a}, \qquad (A.5)$$

and Eq. (A.2) becomes

$$x(t) = k_1 e^{r_1 t} + k_2 e^{r_2 t}. \qquad (A.6)$$

In every instance when the differential equation to be solved has the form of Eq. (A.1), then Eq. (A.6) is the solution. The roots r_1 and r_2 are derived from the circuit components and the constants k_1 and k_2 from the boundary conditions. This can be illustrated by considering two simple circuits involving transients.

Example A.1

Apply a DC voltage V to an RL circuit.
 To calculate the current i (Figure A.1), the differential equation to be solved is

$$V = Ri + L\frac{di}{dt}.$$

Differentiating across gives

$$\frac{d^2 i}{dt^2} + \frac{R}{L}\frac{di}{dt} = 0. \qquad (A.7)$$

This equation has the format of Eq. (A.1), where $a = 1$, $b = R/L$ and $c = 0$. The roots r_1 and r_2 are

$$r_1 = -\frac{R}{2L} + \sqrt{\left(\frac{R}{2L}\right)^2 - (0)}$$
$$= 0$$

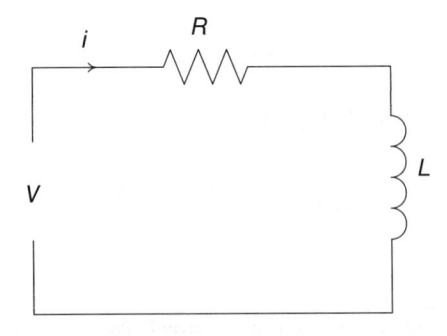

Figure A.1 RL circuit.

Appendix A: Differential Equations

and

$$r_2 = -\frac{R}{2L} - \sqrt{\left(\frac{R}{2L}\right)^2 - (0)}$$
$$= -R/L.$$

From Eq. (A.6), the solution to Eq. (A.7) is

$$i = k_1 + k_2 \, e^{-(R/L)t}.$$

To calculate the constants k_1 and k_2, we need to apply the boundary conditions.
Boundary condition 1: At $t = 0$, $i = 0$.

$$0 = k_1 + k_2 \, e^0,$$

giving $k_2 = -k_1$ and

$$i = k_1 \left(1 - e^{-(R/L)t}\right).$$

Boundary condition 2: At $t = \infty$ or steady state, $i = V/R$.

$$V/R = k_1(1 - e^{-\infty}),$$
$$k_1 = V/R,$$

and the solution for Eq. (A.7) is

$$i = \frac{V}{R}\left(1 - e^{-(R/L)t}\right)$$

or

$$i = I\left(1 - e^{-(R/L)t}\right), \tag{A.8}$$

where I is the steady-state DC current.

Example A.2

Apply a DC voltage V to an RC circuit.
  To calculate the capacitor voltage variation with time, we need first to calculate the current i (Figure A.2).

$$V = Ri + \frac{q}{c}.$$

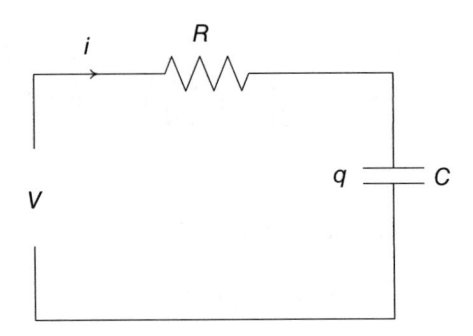

Figure A.2 RC circuit.

Differentiating across,

$$\frac{di}{dt} + \frac{1}{RC} \cdot \frac{dq}{dt} = 0$$

or

$$\frac{di}{dt} + \frac{1}{RC} i = 0.$$

Differentiating again,

$$\frac{d^2 i}{dt^2} + \frac{1}{RC} \frac{di}{dt} = 0. \tag{A.9}$$

This equation also has the format of Eq. (A.1), where $a = 1$, $b = 1/RC$ and $c = 0$. The roots r_1 and r_2 are

$$r_1 = -\frac{1}{2RC} + \sqrt{\left(\frac{1}{2RC}\right)^2 - 0}$$
$$= 0,$$
$$r_2 = -\frac{1}{2RC} - \sqrt{\left(\frac{1}{2RC}\right)^2 - 0}$$
$$= -\frac{1}{RC}.$$

From Eq. (A.6), the solution to Eq. (A.9) is

$$i = k_1 + k_2 \, e^{-t/RC}.$$

Boundary condition 1: At $t = 0$, $q = 0$.

$$V = R(k_1 + k_2 \, e^0) + 0,$$
$$k_1 + k_2 = V/R.$$

Appendix A: Differential Equations

Boundary condition 2: At $t=\infty$ or steady state, the capacitor voltage v_c equals V and $i=0$.

$$0 = k_1 + k_2\, e^{-\infty},$$
$$k_1 = 0,$$

and the solution for Eq. (A.9) is

$$i = \frac{V}{R} e^{-t/RC}. \tag{A.10}$$

To now calculate the capacitor voltage v_c,

$$\begin{aligned}
v_c &= \frac{q}{c} \\
&= \frac{1}{c}\int i \cdot dt \\
&= \frac{V}{RC}\int e^{-t/RC} \cdot dt \\
&= \frac{V}{RC}\left[-\frac{1}{1/RC} e^{-t/RC}\right]_0^t \\
&= V\left[-e^{-t/RC}\right]_0^t \\
&= V\left[-e^{-t/RC} - (-e^0)\right] \\
&= V\left(1 - e^{-t/RC}\right).
\end{aligned} \tag{A.11}$$

The transients described by Eqs. (A.8) and (A.11) are exponential with time constants L/R and $1/RC$, respectively. Considering only the term within the brackets, if we set $t = L/R$ or $t = 1/RC$, respectively, we get

$$\left(1 - e^{-1}\right) = 0.632.$$

The time constants are thus the time it takes for either the current i (Eq. (A.8)) or the voltage v_c (Eq. (A.11)) to reach 0.632 pu or their respective final values.

Returning to Eqs. (A.4) and (A.5), three cases are possible dependent on the relative values of the terms under the square root sign, that is, b^2 and $4ac$.

Case 1: $b^2 - 4ac > 0$ and the roots are real.
Rewriting Eqs. (A.4) and (A.5),

$$r_1 = -\frac{b}{2a} + \sqrt{\left(\frac{b}{2a}\right)^2 - \frac{c}{a}},$$

$$r_2 = -\frac{b}{2a} - \sqrt{\left(\frac{b}{2a}\right)^2 - \frac{c}{a}}.$$

Let
$$\alpha = \frac{b}{2a}$$

and

$$\beta = \sqrt{\left(\frac{b}{2a}\right)^2 - \frac{c}{a}}$$

and the roots become

$$r_1 = -\alpha + \beta,$$
$$r_2 = -\alpha - \beta.$$

Equation (A.6) is now rewritten as

$$\begin{aligned} x(t) &= k_1 \, e^{(-\alpha+\beta)t} + k_2 \, e^{(-\alpha-\beta)t} \\ &= e^{-\alpha t}\left(k_1 \, e^{\beta t} + k_2 \, e^{-\beta t}\right). \end{aligned} \quad (A.12)$$

Equation (A.12) could be left in this format, but a more flexible approach is to convert it to hyperbolic functions. In general,

$$e^x = \cosh x + \sinh x,$$
$$e^{-x} = \cosh x - \sinh x.$$

Equation (A.12) becomes

$$\begin{aligned} x(t) &= e^{-\alpha t}[k_1(\cosh \beta t + \sinh \beta t) + k_2(\cosh \beta t - \sinh \beta t)] \\ &= e^{-\alpha t}[(k_1 + k_2)\cosh \beta t + (k_1 - k_2)\sinh \beta t] \\ &= e^{-\alpha t}[k_3 \cosh \beta t + k_4 \sinh \beta t]. \end{aligned} \quad (A.13)$$

Equation (A.13) corresponds to the overdamped cases in RLC oscillation circuit theory.
Case 2: $b^2 = 4ac$ and the roots are equal and real.

$$r_1 = r_2 = r \quad \text{and} \quad r = -\alpha.$$

$x(t) = k \, e^{rt}$ cannot be a solution because the solution to a second-order differential equation must have two arbitrary constants. We can therefore write

$$x(t) = y \, e^{rt},$$

where y is a function of t and has two constants. The simplest possible equation for y is

$$y = k_1 + k_2 t$$

Appendix A: Differential Equations

and

$$x(t) = (k_1 + k_2 t) e^{rt}. \tag{A.14}$$

We can now test if Eq. (A.14) is a solution for Eq. (A.1). Note first that

$$r = -\frac{b}{2a}$$

or $2ar + b = 0$,

$$\begin{aligned}
a\frac{d^2 x(t)}{dt^2} &+ b\frac{dx(t)}{dt} + cx(t) \\
&= a\left[k_1 r^2 e^{rt} + k_2 r e^{rt} + k_2 e^{rt} + k_2 e^{rt}(r^2 t + r)\right] \\
&\quad + b[k_1 r e^{rt} + k_2 e^{rt} + k_2 r t e^{rt}] \\
&\quad + c[k_1 e^{rt} + k_2 t e^{rt}] \\
&= k_1 e^{rt}\left[ar^2 + br + c\right] + k_2 e^{rt}[2ar + b] \\
&\quad + k_2 t e^{rt}\left[ar^2 + br + c\right] \\
&= 0.
\end{aligned}$$

Note: r is the root of the auxiliary equation (Eq. (A.3)), which is equal to zero. Equation (A.14) is therefore a solution for this case, and substituting for r, the general equation is

$$x(t) = (k_1 + k_2 t) e^{-\alpha t}. \tag{A.15}$$

Equation (A.15) corresponds to the critically damped case in RLC oscillation circuit theory.

Case 3: $b^2 < 4ac$ and the roots are complex.
The roots are

$$r_1 = -\frac{b}{2a} + \sqrt{(-1)\left[\frac{c}{a} - \left(\frac{b}{2a}\right)^2\right]}$$

$$= -\alpha + j\beta, \quad \text{where } j = \sqrt{-1},$$

$$r_2 = -\alpha - j\beta.$$

Note: β now is

$$\sqrt{\frac{c}{a} - \left(\frac{b}{2a}\right)^2}.$$

The solution for Eq. (A.1) is

$$x(t) = e^{-\alpha t}\left(k_1 e^{j\beta t} + k_2 e^{-j\beta t}\right). \tag{A.16}$$

Table A.1 Solutions for $a(d^2x/dt^2) + b(dx/dt) + cx = 0$.

Case	Equation	Solution
$b^2 > 4ac$	(A.13)	$x(t) = e^{-\alpha t}(k_1 \cosh \beta t + k_2 \sinh \beta t)$
$b^2 = 4ac$	(A.15)	$x(t) = (k_1 + k_2 t)e^{-\alpha t}$
$b^2 < 4ac$	(A.17)	$x(t) = e^{-\alpha t}(k_1 \cos \beta t + k_2 \sin \beta t)$

Euler's equations are

$$e^{jx} = \cos x + j \sin x,$$
$$e^{-jx} = \cos x - j \sin x,$$

and substituting in Eq. (A.16)

$$\begin{aligned} x(t) &= e^{-\alpha t}[k_1(\cos \beta t + j \sin \beta t) + k_2(\cos \beta t - j \sin \beta t)] \\ &= e^{-\alpha t}[(k_1 + k_2)\cos \beta t + j(k_1 - k_2)\sin \beta t] \qquad (A.17)\\ &= e^{-\alpha t}[k_3 \cos \beta t + k_4 \sin \beta t]. \end{aligned}$$

Equation (A.17) corresponds to the underdamped case in RLC oscillation circuit theory.

We can summarize the three possible solutions for Eq. (A.1) as shown in Table A.1, where k_1 and k_2 are arbitrary constants dependent on the boundary conditions.

The equations in Table A.1 can be used to calculate the current interruption transients associated with the circuits (a), (b) and (c) in Figure 2.1.

The second type of differential equation that is applicable is the second-order non-homogenous linear differential equation, which takes the form

$$a\frac{d^2x}{dt^2} + b\frac{dx}{dt} + cx = F(x). \qquad (A.18)$$

The general solution for equations of this type can be written as

$$x(t) = x_p(t) + x_c(t), \qquad (A.19)$$

where $x_p(t)$ is the particular solution of Eq. (A.18) and $x_c(t)$ is the solution of the related homogenous equation

$$a\frac{d^2x}{dt^2} + b\frac{dx}{dt} + cx = 0, \qquad (A.20)$$

which is, of course, identical to Eq. (A.1) and now referred to as the complementary equation.

This type of differential equation is applicable to a variation of circuit (c) in Figure 2.1 and directly to circuit (d). Such equations are solved by using the method of undetermined

Appendix A: Differential Equations

Table A.2 Perspective solutions for non-homogenous second-order differential equations.

If $F(x)$ is	And	Use
A polynomial $ax^2 + bx + c$ of degree 2	Zero is not a characteristic root	A polynomial $Ax^2 + Bx + C$
$\sin kx$ of $\cos kx$	kj is not a characteristic root	$A \cos kx + B \sin kx$
e^{nx}	n is a single characteristic root	$Cx e^{nx}$

coefficients. Basically, a guess is made for the solution dependent on the nature of $F(x)$. If $F(x)$ is a polynomial of a certain degree, an exponential function or a sinusoidal function, then the guesses would be similar functions but with coefficients to be determined. Textbooks on differential equations provide lookup tables of the prospective solutions such as shown in Table A.2.

For example, if $F(x) = x^2$,

$$a\frac{d^2x}{dt^2} + b\frac{dx}{dt} + cx = x^2. \tag{A.21}$$

The solution is taken to be a polynomial of the second degree (same power as $F(x)$), giving

$$x_p(t) = Ax^2 + Bx + C,$$

where A, B and C are the as-yet undetermined coefficients.

$$\frac{dx_p(t)}{dt} = 2Ax + B,$$

$$\frac{d^2 x_p(t)}{dt^2} = 2A.$$

Substituting into Eq. (A.21),

$$a(2A) + b(2Ax + B) + c(Ax^2 + Bx + C) = x^2,$$
$$(cA)x^2 + (2Ab + Bc)x + 2aA + bB + cC = x^2,$$

giving

$$cA = 1, \quad 2Ab + Bc = 0, \quad 2aA + bB + cC = 0.$$

a, b and c are known, and A, B and C can be determined.

If $F(x)$ is a constant K, then the solution is a polynomial of zero degree:

$$x_p(t) = C.$$

From Eq. (A.21), we write

$$a(0) + b(0) + c \cdot C = K,$$

giving

$$C = \frac{K}{c}.$$

The solution is then of the form

$$x(t) = k_1 e^{r_1 t} + k_2 e^{r_2 t} + K/c. \tag{A.22}$$

If $F(x)$ is a sinusoidal function $K \sin \omega t$ or $K \cos \omega t$, then $x_p(t)$ is taken as

$$x_p(t) = A \cos \omega t + B \sin \omega t.$$

From Eq. (A.18),

$$a(-A\omega^2 \cos \omega t - B\omega^2 \sin \omega t) + b(-A\omega \sin \omega t + B\omega \cos \omega t) + c(A \cos \omega t + B \sin \omega t)$$
$$= K \sin \omega t \text{ or } K \cos \omega t.$$

For $F(x) = K \sin \omega t$, we can write

$$-aA\omega^2 + bB\omega + Ac = 0, \tag{A.23}$$

$$-aB\omega^2 - bA\omega + cB = K. \tag{A.24}$$

For $F(x) = K \cos \omega t$, the quantities on the right are reversed to give

$$-aA\omega^2 + bB\omega + Ac = K, \tag{A.25}$$

$$-aB\omega^2 - bA\omega + cB = 0. \tag{A.26}$$

a, b, c and K are known, and A and B can be determined using either set of equations. Non-homogeneous differential equations in which $F(x)$ is a sinusoidal function are typically used in solving steady-state AC circuits.

Bibliography

1. Stewart, J. (2009) *Calculus*, Thomson Brooks/Cole.

Appendix B
Principle of Duality

In electrical engineering, a dual of a relationship is formed by interchanging voltage and current in an expression. The notion is more complex than that because reciprocal circuit elements also need to be considered. Basic electrical dualities are shown in Table B.1.

A simple example is Ohm's law, where we can write $v = Ri$ and apply the principle of duality to derive $i = Gv$.

Duality means that the current or voltage in one circuit behaves in a similar manner as the voltage or current, respectively, in another circuit. For example, if we apply a DC voltage V to a series RL circuit, then the equation to derive the current i is

$$V = Ri + L\frac{di}{dt},$$

and the solution is

$$i = \frac{E}{R}\left(1 - e^{-(R/L)t}\right).$$

Take now a parallel RC circuit and inject a current I for which the corresponding equation and solution for the common voltage v are

$$I = \frac{v}{R} + C\frac{dv}{dt}$$

and

$$v = RI\left(1 - e^{-(1/RC)t}\right).$$

Clearly, by making the duality conversions given in Table B.1, either set of equations can be derived from the other, and duality exists between the two cases.

Duality also exists between series and parallel RLC circuits, but care needs to be exercised in applying it. Duality does not exist between the degrees of damping d_s and d_p as is evident from

---

Current Interruption Transients Calculation, First Edition. David F. Peelo.
© 2014 John Wiley & Sons, Ltd. Published 2014 by John Wiley & Sons, Ltd.

Table B.1 Electrical dualities.

	Duality between	
Voltage	↔	Current
Series	↔	Parallel
Resistance	↔	Conductance
Capacitance	↔	Inductance
Reactance	↔	Susceptance
Short circuit	↔	Open circuit
KCL[a]	↔	KVL[b]

[a] Kirchoff's current law.
[b] Kirchoff's voltage law.

Table B.2 Comparison of series and parallel RLC circuits with respect to degree of damping.

Oscillation	Series RLC, $d_s = \frac{R}{R_C}$	Parallel RLC, $d_p = \frac{R_C}{R}$
Overdamped	$R > R_C$	$R < R_C$
Critically damped	$R = R_C = 2\sqrt{\frac{L}{C}}$	$R = R_C = 0.5\sqrt{\frac{L}{C}}$
Underdamped	$R < R_C$	$R > R_C$

Table B.2. d_s is not the reciprocal of d_p because the respective R_C values are not duals. However, the principle can be used bearing in mind this limitation.

In Section 2.5, we considered the case of ramp current injection for parallel circuit shown in Figure 2.7. The differential equation to be solved (Eq. (2.31)) is

$$\frac{d^2v}{dt^2} + \frac{1}{RC}\frac{dv}{dt} + \frac{1}{LC} = \frac{I}{C}, \tag{B.1}$$

with the solution

$$v = LI\left[1 - e^{-\alpha t}\left(\cos \beta t + \frac{\alpha}{\beta}\sin \beta t\right)\right]. \tag{B.2}$$

The dual for this case is the injection of a ramp voltage $V \cdot t$ into the series RLC circuit of Figure 2.1a. We can write

$$Ri + L\frac{di}{dt} + \frac{q}{C} = V \cdot t,$$

which, after differentiation, becomes

$$\frac{d^2i}{dt^2} + \frac{R}{L}\frac{di}{dt} + \frac{1}{LC}i = \frac{V}{L}. \tag{B.3}$$

Appendix B: Principle of Duality

and the dual of Eq. (B.1). The solution for Eq. (B.3) then is

$$i = VC\left[1 - e^{-\alpha t}\left(\cos \beta t + \frac{\alpha}{\beta}\sin \beta t\right)\right].\quad (B.4)$$

The duality conversion is applied also to the values of α and β:

$$\alpha_p = \frac{1}{2RC} \rightarrow \alpha_s = \frac{R}{2L}$$

$$\beta_p = \sqrt{\frac{1}{LC} - \left(\frac{1}{2RC}\right)^2} \rightarrow \beta_s = \sqrt{\frac{1}{LC} - \left(\frac{R}{2L}\right)^2}.$$

The corresponding per-unit generic versions for Eqs. (B.2) and (B.4), respectively, are

$$v_{pu} = 1 - e^{-d_p t_g}\left(\cos\sqrt{1 - d_p^2}\,t_g + \frac{d_p}{\sqrt{1-d_p^2}}\sin\sqrt{1 - d_p^2}\,t_g\right)$$

and

$$i_{pu} = 1 - e^{-d_s t_g}\left(\cos\sqrt{1 - d_s^2}\,t_g + \frac{d_s}{\sqrt{1-d_s^2}}\sin\sqrt{1 - d_s^2}\,t_g\right).$$

These equations have the same format, and a common set of generic curves can be drawn on the basis of d_p and d_s being numerically equal. Note that, from Table B.2, the actual relationship between d_p and d_s for the values of R, L and C is given by

$$\frac{d_p}{d_s} = \frac{R_{Cp}}{R_{Cs}} = 0.25,$$

where R_{Cp} and R_{Cs} are the critical resistance values for the parallel and series cases, respectively.

Appendix C

Useful Formulae

1. Sinusoidal, hyperbolic, exponential and complex functions:

$$\sin^2 x + \cos^2 x = 1,$$

$$\sin(x+y) = \sin x \cos y + \cos x \sin y,$$

$$\sin(x-y) = \sin x \cos y - \cos x \sin y,$$

$$\cos(x+y) = \cos x \cos y - \sin x \sin y,$$

$$\cos(x-y) = \cos x \cos y + \sin x \sin y,$$

$$\left.\begin{array}{l} e^{ix} = \cos x + i \sin x \\ e^{-ix} = \cos x - i \sin x \end{array}\right\} \quad \text{Euler's formulae}, \quad i = \sqrt{-1},$$

$$\sin x = \frac{1}{2i}\left(e^{ix} - e^{-ix}\right),$$

$$\cos x = \frac{1}{2i}\left(e^{ix} + e^{-ix}\right),$$

$$\sinh ix = \frac{1}{2}\left(e^{ix} - e^{-ix}\right),$$

$$\cosh ix = \frac{1}{2}\left(e^{ix} + e^{-ix}\right),$$

$$\sin(x+iy) = \sin x \cosh y + i \cos x \sinh y,$$

$$\cos(x+iy) = \cos x \cosh y - i \sin x \sinh y.$$

Current Interruption Transients Calculation, First Edition. David F. Peelo.
© 2014 John Wiley & Sons, Ltd. Published 2014 by John Wiley & Sons, Ltd.

Appendix C: Useful Formulae

2. Derivatives:

$$\frac{d}{dx}c = 0 \quad (c = \text{constant}),$$

$$\frac{d}{dx}x = 1,$$

$$\frac{d}{dx}cx = c,$$

$$\frac{d}{dx}x^c = cx^{c-1},$$

$$\frac{d}{dx}\left(\frac{1}{x}\right) = -\frac{1}{x^2},$$

$$\frac{d}{dx}\left(\frac{1}{x^c}\right) = -\frac{c}{x^{c+1}},$$

$$\frac{d}{dx}e^x = e^x,$$

$$\frac{d}{dx}e^{g(x)} = \frac{d}{dx}g(x) \cdot e^{g(x)},$$

$$\frac{d}{dx}\ln x = \frac{1}{x},$$

$$\frac{d}{dx}\sin ax = a\cos ax,$$

$$\frac{d}{dx}\cos ax = -a\sin ax,$$

$$\frac{d}{dx}\tan x = \sec^2 x,$$

$$\frac{d}{dx}\sinh ax = a\cosh x,$$

$$\frac{d}{dx}\cosh ax = a\sinh x,$$

$$\frac{d}{dx}uv = u\frac{dv}{dx} + v\frac{du}{dx}.$$

3. Integrals:

$$\int dx = x + c \quad (c = \text{constant}),$$

$$\int a\,dx = ax + c,$$

$$\int x^a\,dx = \frac{1}{a+1}x^{a+1} + c,$$

$$\int \frac{1}{x}\,dx = \ln x,$$

$$\int e^x\,dx = e^x + c,$$

$$\int e^{ax}\,dx = \frac{1}{a}e^{ax} = c,$$

$$\int \sin ax\,dx = -\frac{1}{a}\cos ax + c,$$

$$\int \cos ax\,dx = \frac{1}{a}\sin ax + c,$$

$$\int \sinh ax\,dx = \frac{1}{a}\cosh ax + c,$$

$$\int \cosh ax\,dx = \frac{1}{a}\sinh ax + c,$$

$$\int u\cdot dv = uv - \int v\,du \quad (\text{integration by parts}),$$

$$\int x\sin ax\,dx = \frac{1}{a^2}\sin ax - \frac{x}{a}\cos ax,$$

$$\int x\cos ax\,dx = \frac{1}{a^2}\cos ax + \frac{x}{a}\sin ax,$$

$$\int e^{bx}\sin ax\,dx = \frac{e^{bx}}{a^2+b^2}(b\sin ax - a\cos ax),$$

$$\int e^{bx}\cos ax\,dx = \frac{e^{bx}}{a^2+b^2}(a\sin ax + b\cos ax).$$

Appendix D

Euler's Formula

Leonhard Euler (1707–1783) was a prolific Swiss mathematician. His formula[1]

$$e^{ix} = \cos x + i \sin x$$

describes the relationship between the oscillatory trigonometric functions sine and cosine and the complex exponential function where e is the basis of natural logarithms equalling 2.718 2818 and $i = \sqrt{-1}$. e is known as Euler's constant.

The relationship may seem at odds because e^x is a function that increases exponentially with increasing x, and sine and cosine functions oscillate repetitively about an axis. Euler discovered the relationship by considering e^x as a power series and Taylor's series for $\sin x$ and $\cos x$:

$$e^x = \sum_{n=0}^{\infty} \frac{x^n}{n!}$$

$$= 1 + x + \frac{x^2}{2!} + \frac{x^3}{3!} + \frac{x^4}{4!} + \frac{x^5}{5!} + \cdots.$$

Substituting i^x for x,

$$e^{ix} = 1 + ix - \frac{x^2}{2!} - i\frac{x^3}{3!} + \frac{x^4}{4!} + i\frac{x^5}{5!} - \frac{x^6}{6!} - i\frac{x^7}{7!} + \frac{x^8}{8!} + \cdots$$

$$= \left(1 - \frac{x^2}{2!} + \frac{x^4}{4!} - \frac{x^6}{6!} + \cdots\right) + i\left(x - \frac{x^3}{3!} + \frac{x^5}{5!} - \frac{x^7}{7!} + \cdots\right)$$

$$= \cos x + i \sin x,$$

the quantities in brackets being Taylor's series for $\cos x$ and $\sin x$, respectively.

---

[1] Mathematicians use the notation i for the imaginary number $\sqrt{-1}$. Electrical engineers use the symbol j for $\sqrt{-1}$ because i is reserved for current.

Current Interruption Transients Calculation, First Edition. David F. Peelo.
© 2014 John Wiley & Sons, Ltd. Published 2014 by John Wiley & Sons, Ltd.

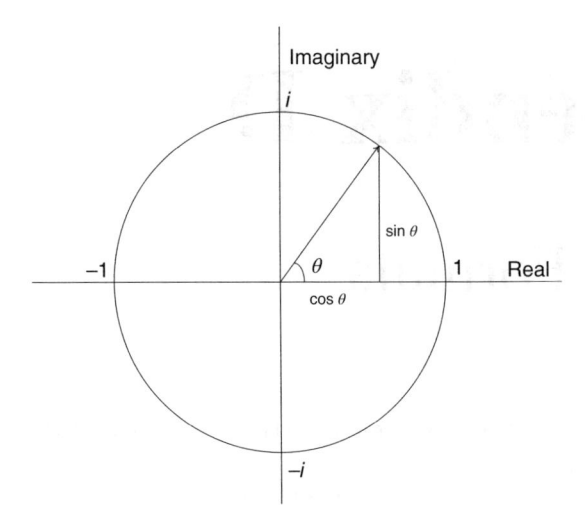

Figure D.1 Euler formula plot.

Although Euler may not have recognized it, the formula is the basis for complex number theory and links Cartesian and polar coordinates. We can write

$$\begin{aligned} z &= x + iy \\ &= |z|(\cos\theta + i\sin\theta) \\ &= r\,e^{j\theta}, \end{aligned}$$

where x is the real part, y is the imaginary part and

$$|z| = \sqrt{x^2 + y^2} = r,$$
$$\cos\theta = \frac{x}{r} \quad \text{and} \quad \sin\theta = \frac{y}{r}.$$

Taking the formula as a whole, it represents a unit circle in a complex plane with a vector as radius as shown in Figure D.1. If we take $\theta = 0$,

$$e^{i0} = 1 + i0,$$

and only the real part exists as a horizontal vector. If we now multiply e^{i0} by $e^{i\theta}$, we get

$$e^{i0} \cdot e^{i\theta} = e^{i\theta} = \cos\theta + i\sin\theta,$$

essentially rotating the vector through $\theta°$. Therefore, for the vector at any location around the circuit, multiplication by $e^{i\theta}$ will cause it to rotate $\theta°$ counterclockwise from its original position.

To now consider the right-hand side of the formula, we will view the tip of the vector as it moves on the circumference of the circle. Viewing the tip from the right and in the plane of the

Appendix D: Euler's Formula

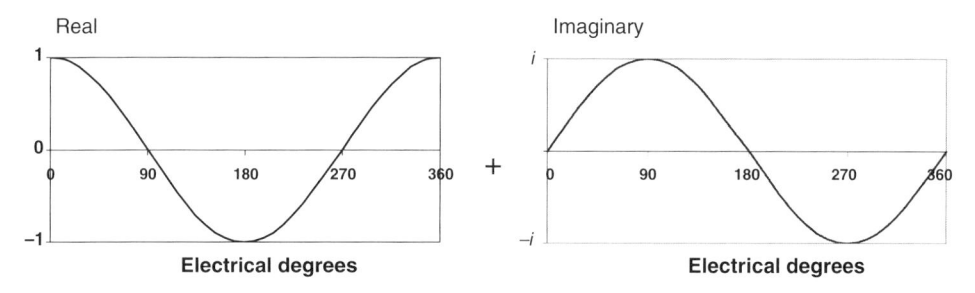

Figure D.2 Euler formula right-hand side plot.

circuit, we can plot the trajectory of the tip for the real and imaginary parts for one complete rotation as shown in Figure D.2. The real part will start at 1, continue to 0 at 90°, −1 at 180°, 0 again at 270° and back to 1 at 360°. Likewise, the imaginary part will start at 0, continue to i at 90° and so on. Figures D.1 and D.2 show the link between the $e^{j\theta}$ rotation on the left-hand side of the formula to the sinusoidal oscillations of the right-hand side.

To introduce an element of time, that is, the rate at which the vector rotates, we replace θ with ωt, where ω is the angular frequency in rad/s, giving

$$e^{ji\omega t} = \cos \omega t + i \sin \omega t.$$

This equation is used to describe steady-state voltages and currents in AC systems. For 50 and 60 Hz systems, ω equals 314 and 377 rad/s, respectively.

For the symmetrical component calculations discussed in Chapter 3, the rotation of interest is through 120° based on angle between phases in a three-phase AC system. We now introduce the operator a, which gives this rotation, and reverting to electrical engineering notation, we can write

$$\begin{aligned} a &= e^{j120} \\ &= \cos 120° + j \sin 120° \\ &= -0.5 + j\,0.866, \end{aligned}$$

often written as

$$a = -\frac{1}{2} + j\sqrt{\frac{3}{2}}.$$

For a 240° rotation,

$$\begin{aligned} a^2 &= a.a \\ &= e^{j120} \cdot e^{j120} \\ &= e^{j240} \\ &= \cos 240° + j \sin 240° \\ &= -0.5 - j0.866. \end{aligned}$$

As a final note, if we take $\theta = \pi$,

$$\begin{aligned} e^{i\pi} &= \cos\pi + j\sin\pi \\ &= -1 + j0 \end{aligned}$$

or

$$e^{i\pi} + 1 = 0.$$

This remarkable equation, known as Euler's identity, shows the link between the mathematical constants e, i, π, 0 and 1.

Bibliography

1. Maor, E. (1994) *e: The Story of a Number*, Princeton University Press.

Appendix E

Asymmetrical Current-Calculating Areas Under Curves

A maximum asymmetrical curve is given by Eq. (4.26), which, to account for polarity, can be written as

$$i_{max} = \pm \sqrt{2}I \left(\cos \omega t - e^{-t/\tau} \right),$$

where I is the circuit breaker short-circuit current rating and τ is the system DC time constant. The area under any loop of current can be obtained by integration of i_{max} between the two applicable current zero crossing times t_1 and t_2:

$$\text{Loop area} = \pm I\sqrt{2}I \int_{t_1}^{t_2} \left(\cos \omega t - e^{-t/\tau} \right) dt$$
$$= \pm I\sqrt{2}I \left[\frac{1}{\omega} \sin \omega t + \tau e^{-t/\tau} \right]_{t_1}^{t_2}. \quad (E.1)$$

Taking the case of 63 kA, 50 Hz and $\tau = 45$ ms and the major loop shown in Figure 4.37, t_1 and t_2 are 23 and 36.7 ms, respectively. Using the negative version of Eq. (E.1), readers can readily determine that the loop area is 1.105 kA s.

Having plotted the asymmetrical current as shown for the example in Figure 4.35, the area under any particular loop of current can be estimated using the so-called trapezoidal rule. With reference to Figure E.1, any area under a curve can be split up into a series of trapezoids of width t_s, and the area A_{TR} under the curve is the combined areas of the individual trapezoids:

$$A_{TR} = t_s \left(\frac{i_0 + i_1}{2} \right) + t_s \left(\frac{i_1 + i_2}{2} \right) + \cdots + t_s \left(\frac{i_{n-1} + i_n}{2} \right)$$
$$= \frac{t_s}{2} (i_0 + 2i_1 + 2i_2 + \cdots + 2i_{n-1} + i_n). \quad (E.2)$$

Current Interruption Transients Calculation, First Edition. David F. Peelo.
© 2014 John Wiley & Sons, Ltd. Published 2014 by John Wiley & Sons, Ltd.

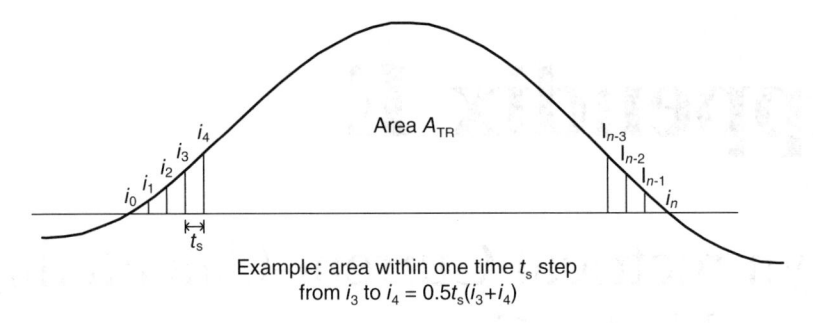

Figure E.1 Application of the trapezoidal rule.

Example: area within one time t_s step from i_3 to $i_4 = 0.5t_s(i_3+i_4)$

A second and more subtle approach is the use of Simpson's rule. Simpson proposed that the area under a curve could be approximated by means of the parabolic equation as shown in Figure E.2. We can write

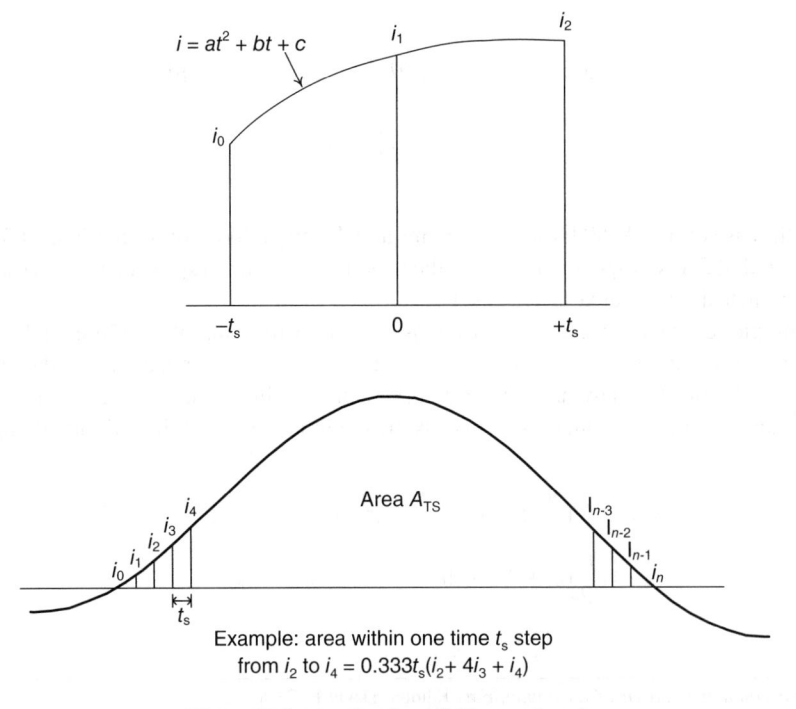

$$i = at^2 + bt + c, \tag{E.3}$$

Example: area within one time t_s step from i_2 to $i_4 = 0.333t_s(i_2 + 4i_3 + i_4)$

Figure E.2 Application of Simpson's rule.

Appendix E: Asymmetrical Current-Calculating Areas Under Curves

and the area A under the curve is given by

$$A = \int_{-t_s}^{t_s} (at^2 + bt + c) \cdot dt$$

$$= \left[\frac{at^3}{3} + \frac{bt^2}{2} + ct\right]_{-t_s}^{t_s}$$

$$= \frac{t_s}{3}\left[2at_s^2 + bc\right].$$

From Eq. (E.3) and Figure E.2,

$$i_0 = at_s^2 - bt_s + c,$$
$$i_1 = c,$$
$$i_2 = at_s^2 + bt_s + c$$

and (as discovered by Simpson)

$$i_0 + 4i_1 + i_2 = \left(at_s^2 - bt_s + c\right) + 4c + \left(at_s^2 + bt_s + c\right)$$

$$= 2at_s^2 + bc,$$

giving

$$A = \frac{t_s}{3}(i_0 + 4i_1 + i_2). \tag{E.4}$$

Equation (E.4) can be extended to a complete curve of total area A_{SR} (Figure E.2):

$$A_{SR} = \frac{t_s}{3}(i_0 + 4i_1 + i_2) + \frac{t_s}{3}(i_2 + 4i_3 + i_4) + \cdots + \frac{t_s}{3}(i_{n-2} + 4i_{n-1} + i_n)$$

$$= \frac{t_s}{3}(i_0 + 4i_1 + 2i_2 + 4i_3 + 2i_4 + \cdots + 2i_{n-2} + 4i_{n-1} + i_n). \tag{E.5}$$

Applying Eqs. (E.2) and (E.5) to the major loop of Figure 4.35 with a time step $t_s = 0.2$ ms gives the same result of 1.104 kA s. For those who wish to calculate loop areas more accurately than the suggested comparison approach in IEC 62271-100, either method is acceptable.

Appendix F

Shunt Reactor Switching: First-Pole-to-Clear Circuit Representation

A shunt reactor, quite apart from its function, is an inductive load on the system. The shunt reactor Mvar rating is generally very much less than the system MVA short-circuit level at its location. For the overvoltage calculations of Chapter 6, the system was therefore treated as being "infinite," and the purpose of this appendix is to justify that approach using symmetrical components.

The circuit for the general shunt reactor switching case—or, in fact, any balanced load with neutral impedance earthing—is shown in Figure F.1. Each shunt reactor phase is represented by an impedance Z, and the neutral is earthed through an impedance Z_e. V_a, V_b and V_c are phase voltages to earth, and V_e is the neutral voltage to earth. Before first pole clearing, the circuit is balanced and V_e is floating at earth potential. After current interruption in a-phase, we can write the "fault" equations as

$$I_a = 0,$$
$$V_b = I_b(Z + Z_e) + I_c Z_e,$$
$$V_c = I_c(Z + Z_e) + I_b Z_e,$$
$$V_e = (I_b + I_c)Z_e,$$

and V_a is the voltage we wish to calculate.

Calculating first the sequence currents,

$$\begin{bmatrix} I_0 \\ I_1 \\ I_2 \end{bmatrix} = T \begin{bmatrix} 0 \\ I_b \\ I_c \end{bmatrix},$$

Current Interruption Transients Calculation, First Edition. David F. Peelo.
© 2014 John Wiley & Sons, Ltd. Published 2014 by John Wiley & Sons, Ltd.

Appendix F: Shunt Reactor Switching

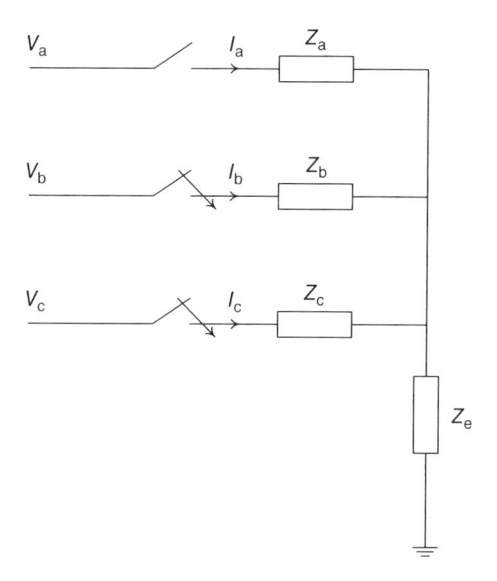

Figure F.1 General circuit for shunt reactor switching calculation.

giving

$$I_0 = \frac{1}{3}(I_b + I_c),$$

$$I_1 = \frac{1}{3}(aI_b + a^2 I_c),$$

$$I_2 = \frac{1}{3}(a^2 I_b + aI_c),$$

$$I_0 + I_1 + I_2 = 0,$$

showing that the sequence networks are connected in parallel.

To find the connection between the positive- and negative-sequence networks, we need to refer both to a common parameter without any zero-sequence component. The common parameter is $V_b - V_c$:

$$\begin{aligned}
V_b - V_c &= (I_b - I_c)(Z + Z_e) + (I_c - I_b)Z_e \\
&= (I_b - I_c)Z \\
&= [(I_0 + a^2 I_1 + aI_2) - (I_0 + aI_1 + a^2 I_2)]Z \\
&= [(a^2 - a)(I_1 - I_2)]Z \\
&= -j\sqrt{3}(I_1 - I_2)Z.
\end{aligned}$$

Expressing now $V_b - V_c$ in terms of the sequences voltages,

$$\begin{aligned}
V_b - V_c &= (V_0 + a^2 V_1 + aV_2) - (V_0 + aV_1 + a^2 V_2) \\
&= (a^2 - a)(V_1 - V_2) \\
&= -j\sqrt{3}(V_1 - V_2) \\
&= -j\sqrt{3}(I_1 - I_2)Z,
\end{aligned}$$

which after rearranging gives
$$V_1 - ZI_1 = V_2 - ZI_2, \tag{F.1}$$

where $V_1 = V_{af} - Z_1 I_1$ and $V_2 = -ZI_2$.

Equation (F.1) confirms the parallel connection and further that the impedance Z is connected in series with each of the sequence networks.

To now calculate the connection of the zero-sequence network, the applicable common parameter is $V_b + V_c$:

$$\begin{aligned} V_b + V_c &= (Z + 2Z_e)(I_b + I_c) \\ &= (Z + 2Z_e)(2I_0 - I_1 - I_2). \end{aligned}$$

Likewise,
$$V_b + V_c = 2V_0 - V_1 - V_2,$$

and we can write
$$2V_0 - V_1 - V_2 = (Z + 2Z_e)(2I_0 - I_1 - I_2).$$

Rearranging gives
$$2V_0 - 2I_0 Z - 4I_0 Z_e = (V_1 - I_1 Z_1) + (V_2 - I_2 Z_2) - Z(I_1 + I_2)Z_e.$$

$I_1 + I_2 = -I_0$ and from Eq. (F.1),
$$2V_0 - 2I_0 Z - 6I_0 Z_e = 2(V_1 - I_1 Z_1)$$

or
$$2V_0 - 2I_0 Z - 6I_0 Z_e = 2(V_2 - I_2 Z_2).$$

Therefore,
$$\begin{aligned} V_0 - I_0 Z - 3I_0 Z_e &= V_1 - I_1 Z_1 \\ &= V_2 - I_2 Z_2. \end{aligned}$$

The negative-sequence network thus has the impedance Z and three times the impedance Z_e connected in series with it. The overall sequence network circuit is shown in Figure F.2.

The sequence currents are
$$I_1 = \frac{V_{af}(2Z + 3Z_e + Z_2 + Z_0)}{(Z_1 + Z)(2Z + 3Z_e + Z_2 + Z_0) + (Z_2 + Z)(Z_0 + 3Z_e + Z)},$$

$$I_2 = \frac{-V_{af}(Z_0 + 3Z_e + Z)}{(Z_1 + Z)(2Z + 3Z_e + Z_2 + Z_0) + (Z_2 + Z)(Z_0 + 3Z_e + Z)},$$

$$I_0 = \frac{-V_{af}(Z_2 + Z)}{(Z_1 + Z)(2Z + 3Z_e + Z_2 + Z_0) + (Z_2 + Z)(Z_0 + 3Z_e + Z)}$$

Appendix F: Shunt Reactor Switching

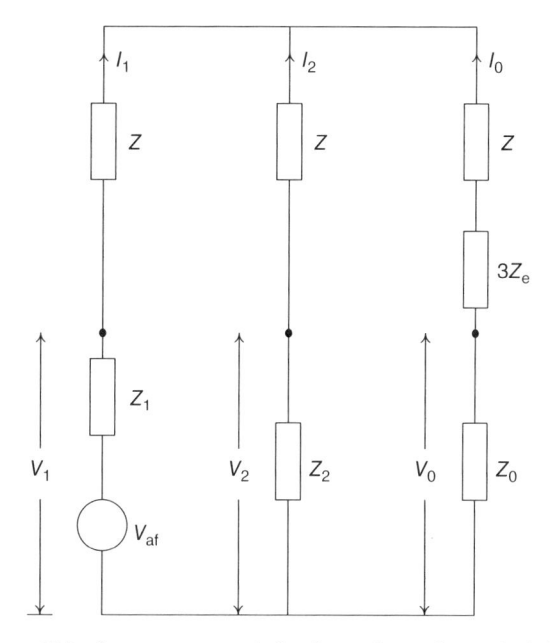

Figure F.2 Sequence network for first-pole-to-clear calculation.

and

$$V_a = V_1 + V_2 + V_0$$
$$= V_{af} - Z_1 I_1 - Z_2 I_2 - Z_0 I_0,$$

giving the first-pole-to-clear factor

$$\frac{V_a}{V_{af}} = \frac{3[Z_0 Z_2 + 2Z_e Z_2 + Z(Z + Z_2 + Z_0 + 2Z_e)]}{(Z_1 + Z)(2Z + 3Z_e + Z_2 + Z_0) + (Z_2 + Z)(Z_0 + 3Z_e + Z)}. \quad \text{(F.2)}$$

To calculate the voltage V_e at the neutral point,

$$V_e = (I_b + I_c) Z_e$$
$$= [2I_0 - (I_1 + I_2)] Z_e$$
$$= [2I_0 - (-I_0)] Z_e$$
$$= 3 I_0 Z_e.$$

The first-pole-to-clear factor for the neutral then is

$$\frac{V_e}{V_{af}} = \frac{-3(Z_2 + Z) Z_e}{(Z_1 + Z)(2Z + 3Z_e + Z_2 + Z_0) + (Z_2 + Z)(Z_0 + 3Z_e + Z)}. \quad \text{(F.3)}$$

As was the practice in the symmetrical component calculations in Chapter 3, we will ignore the resistive components of Z_1, Z_2 and Z_0 and take $X_1 = X_2$. Equations (F.2) and (F.3) are now rewritten as

$$\frac{V_a}{V_{af}} = \frac{3[X_0 X_1 + 2Z_e X_1 + Z(Z + X_1 + X_0 + 2Z_e)]}{(X_1 + Z)(3Z + X_1 + 2X_0 + 6Z_e)}, \quad (F.4)$$

$$\frac{V_e}{V_{af}} = \frac{-3(X_1 + Z)Z_e}{(X_1 + Z)(3Z + X_1 + 2X_0 + 6Z_e)}. \quad (F.5)$$

For an infinite and balanced system, the sequence impedances all equal zero, and Eq. (F.4) becomes

$$\frac{V_a}{V_{af}} = \frac{3[Z(Z + 2Z_e)]}{Z(2Z + 3Z_e) + Z(3Z_e + Z)}$$

$$= 1.$$

Likewise, for the neutral point (Eq. (F.5)),

$$\frac{V_e}{V_{af}} = -\frac{3ZZ_e}{Z(2Z + 3Z_e) + Z(3Z_e + Z)}$$

$$= -\frac{ZZ_e}{Z^2 + 2Z_e Z} \quad (F.6)$$

$$= -\frac{1}{2 + Z/Z_e}.$$

Equation (F.6) defines a ratio of the effective neutral impedance to the impedance Z for the first-pole-to-clear representation. As shown in Figure F.3, the impedance Z in b-phase and c-phase must be connected in parallel with the neutral impedance Z_e. Readers will recognize that Eq. (F.6) gives the neutral shift factor K (Chapter 6), and clearly the applied voltage must equal $V(1 + K)$ to achieve the correct current, where V is the phase voltage peak value to earth. The first-pole-to-clear circuit is that shown in Figure 6.4. For $Z_e = 0$, $K = 1$ and for $Z_e = \infty$, $K = 0.5$.

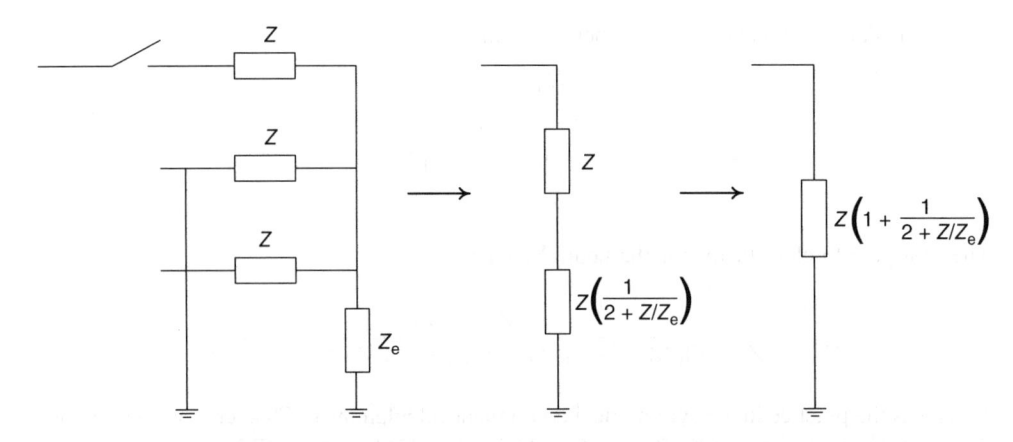

Figure F.3 AC load circuit representation.

Appendix F: Shunt Reactor Switching

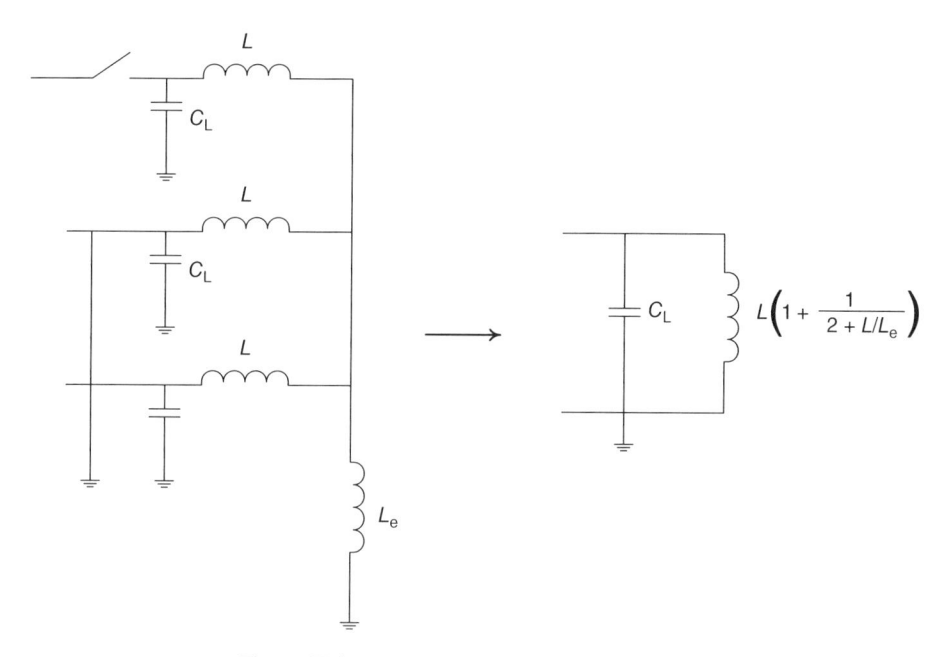

Figure F.4 Transient load circuit representation.

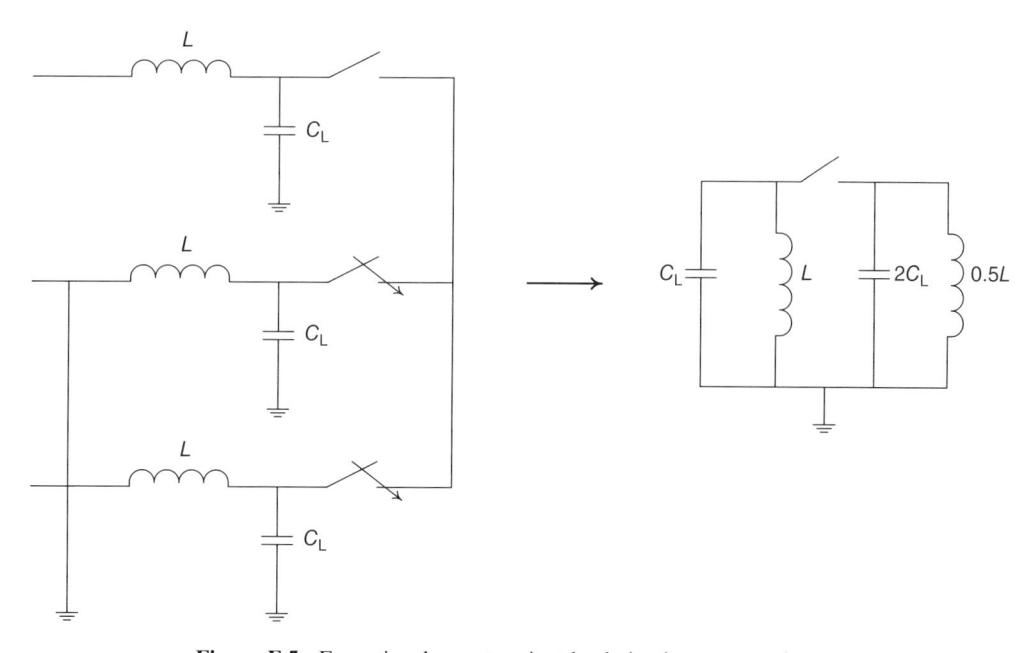

Figure F.5 Exceptional case transient load circuit representation.

Considering the transient circuit and using the L and C_L components, the C_L components in the b-phase and c-phase are shorted out (Figure F.4).

The load circuit will oscillate as one at a frequency of f_{load} given by

$$f_{load} = \frac{1}{2\pi\sqrt{L(1+K)C_L}}.$$

An exceptional isolated neutral case is when the user locates the circuit breaker between the shunt reactor and the neutral (Figure F.5). Transient oscillations will now occur independently on both sides of the circuit but at the same frequency. The source-side oscillation is determined by LC_L, and the neutral-side transient is determined by $0.5L \times 2C_L = LC_L$.

This calculation provides proof of the concept of current injection, that is, shorting out the source and injecting current to derive the circuit response.

Appendix G

Special Case: Interrupting Small Capacitive Currents

It is common practice at all voltage levels to interrupt small capacitive currents up to about 2 A using disconnectors. The disconnectors can be either the air-break type in air-insulated substations (AIS) or SF_6 gas type in gas-insulated substations (GIS). The involved currents are bus charging currents and can also include the currents associated with instrument transformers.

The circuit for this switching duty is shown in Figure G.1. L_S and C_S are the source inductance and the capacitance on the source side of the circuit breaker, respectively, and C_L is the load-side capacitance. As discussed in Chapter 7, interruption of the current leaves a trapped charge Q_L on the load-side capacitance at a voltage V_L, which at least initially equals the peak value V_S of the source voltage. After a half-cycle, the recovery voltage peak across the open disconnector is 2 pu.

A restrike involves two events in sequence. The first event involves the C_S–disconnector–C_L series RLC loop, L being the inductance in the loop, and the restrike causes an equalization of the voltages on C_S and C_L caused by a redistribution of the charges on the two capacitances. We can write

$$Q_S = C_S V_S,$$
$$Q_L = -C_L V_L \quad (V_L \text{ of opposite polarity to } V_S),$$

and assuming a conservation of charge,

$$\begin{aligned} Q_{\text{total}} &= Q_S + Q_L \\ &= (C_S + C_L)V_E, \\ V_E &= \frac{C_S V_S - C_L V_L}{C_S + C_L}. \end{aligned} \tag{G.1}$$

Current Interruption Transients Calculation, First Edition. David F. Peelo.
© 2014 John Wiley & Sons, Ltd. Published 2014 by John Wiley & Sons, Ltd.

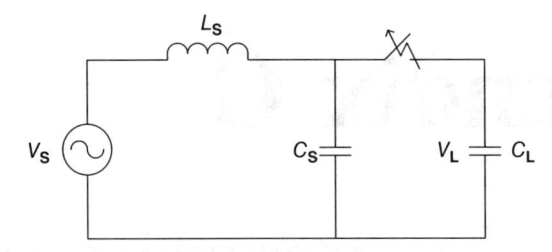

Figure G.1 Capacitive current switching circuit.

Before re-striking, the voltage V_D across the disconnector is given by

$$V_D = V_S - (-V_L)$$
$$= V_S + V_L$$

and substitution in Eq. (G.1) gives

$$V_E = \frac{C_S V_S - C_L(V_D - V_S)}{C_S + C_L}$$
$$= \frac{(C_S + C_L)V_S - C_L V_D}{C_S + C_L}$$
$$= V_S - \frac{C_L}{C_S + C_L} V_D$$

or

$$V_E = V_S - \frac{1}{1 + C_S/C_L} V_D. \qquad (G.2)$$

The oscillation associated with the equalization of the source and load voltages has a starting point at V_L, will reach a peak value dependent on the damping in circuit and has an axis of oscillation at V_E. Equation (G.2) shows that V_E has a certain dependence on the ratio C_S/C_L:

- $C_S > C_L$: the term $(1 + C_S/C_L)$ is large, and V_E will be closer to V_S than to V_L.
- $C_S < C_L$: the term $(1 + C_S/C_L)$ is small, and V_E will be closer to V_L than to V_S.

The second event involves recovery of the voltage V_E to the source voltage V_S. The associated oscillation has a starting point at V_E, an axis of oscillation at V_S and a peak value V_{pk} dependent on the damping in the L_S and parallel C_S and C_L circuit:

$$V_{pk} = V_S + \beta(V_S - V_E)$$
$$= V_S + \beta\left(\frac{V_D}{1 + C_S/C_L}\right), \qquad (G.3)$$

where β is a damping factor and is <1.

Appendix G: Special Case

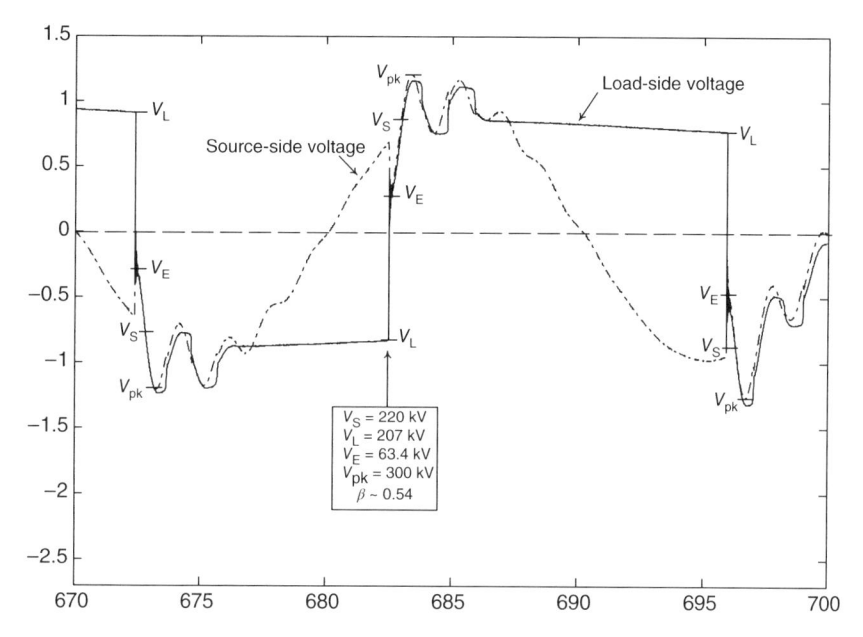

Figure G.2 Capacitive current interruption with 300 kV air-break disconnector and $C_S/C_L = 2.5$ (trace courtesy of KEMA).

The magnitude of the overvoltage V_{pk} also has a dependence on the ratio C_S/C_L:

- $C_S > C_L$: the term $(1 + C_S/C_L)$ is large, and V_{pk} is low. This case is typical for GIS disconnectors, and the repetitive break–make transients are referred to as very fast transients (VFTs).
- $C_S < C_L$: the term $(1 + C_S/C_L)$ is small, and V_{pk} is high. This case is common in AIS installations usually exhibiting long arcing times caused by the energy injected into the arc by the re-striking (inrush) current from the system.

Traces from an actual laboratory test on a 300 kV air-break disconnector are shown in Figures G.2 and G.3.

In Figure G.2, $C_S/C_L = 2.5$, and at the instant of the second restrike in the trace, $V_S = 171$ kV and $V_L = 207$ kV, giving $V_E = 10$ kV (from Eq. (G.2)). This is a favourable C_S/C_L case, and it can be seen that the following power frequency is interrupted as indicated by the load-side DC voltage.

In Figure G.3, $C_S/C_L = 0.04$, and at the instant of the first restrike, $V_S = -202$ kV, $V_L = -154$ kV and $V_E = -140$ kV (from Eq. (G.2)). This is the unfavourable C_S/C_L case, and the value of V_{pk} is much higher than that in Figure G.2. After the first restrike, power frequency current is conducted for a half-cycle and is interrupted at the next current zero with a restrike a further half-cycle later.

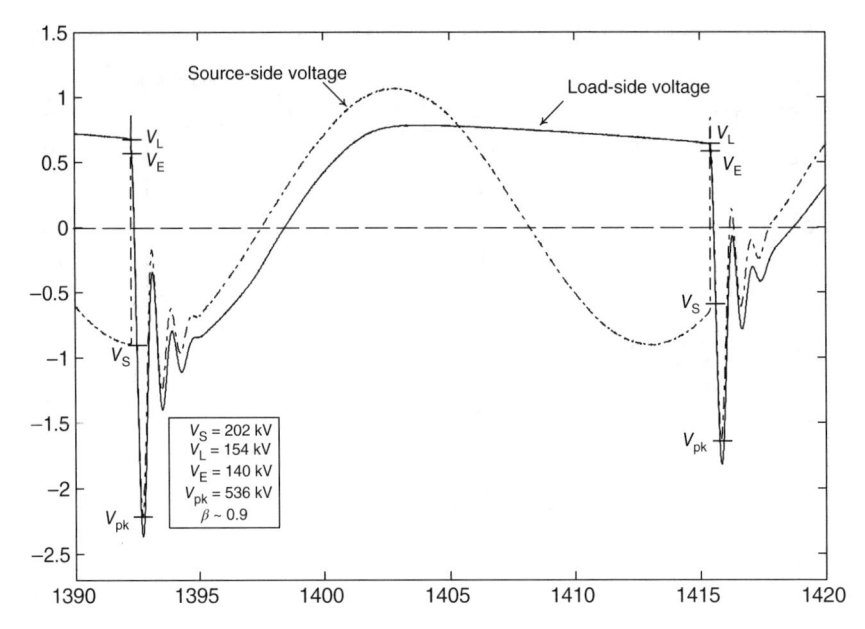

Figure G.3 Capacitive current interruption with 300 kV air-break disconnector and $C_S/C_L = 0.04$ (trace courtesy of KEMA).

Bibliography

1. Peelo, D.F. (2004) Current interruption using high voltage air-break disconnectors. PhD thesis, Eindhoven University of Technology.
2. Peelo, D.F., Smeets, R.P.P., Kuivenhoven, S. and Krone, J.G. (2005) Capacitive current interruption in atmospheric air. Paper No. 106, CIGRE A3 and B3 Colloquium, Tokyo.
3. Chai, Y. (2012) Capacitive current interruption with high voltage air-break disconnectors. PhD thesis, Eindhoven University of Technology.

Appendix H

Evolution of Transient Recovery Voltages

H.1 Introduction

Circuit breakers are one of the greatest inventions of the twentieth century. The first bulk-oil type circuit breaker made longer distance transmission of electric power possible, and development has continued, enabling the development of power systems up to 1200 kV. The approximate manufacturing periods for the different circuit breaker types are as follows:

- *Bulk-oil circuit breakers*: Manufactured from 1910 to 1995, the longest such period for any circuit breaker type.
- *Minimum-oil (also described as low-oil-content) circuit breakers*: Manufactured from 1965 to 1985, a live-tank version of the dead-tank bulk-oil circuit breaker and used extensively outside the United States.
- *Air-blast circuit breakers*: Manufactured from 1950 to 1980, achieving the faster interrupting times needed for the first extra-high-voltage (EHV) power systems at 300 kV and above.
- *Dual-pressure SF_6 gas circuit breakers*: Manufactured from 1955 to 1980, applied principally in the United States.
- *Single-pressure SF_6 gas circuit breakers*: Manufactured from 1965 and continuing, evolving from the straight puffer type to assisted-blast type.
- *Vacuum circuit breakers*: Manufactured from 1965 and continuing, initially at system voltages of 36 kV and below and presently up to system voltages up to 145 kV.

Standardized transient recovery voltages (TRVs) and associated testing have evolved driven by the interrupting medium technology and the growth and knowledge of power systems. The interaction of TRVs with breaking technology was not always known beforehand and standards were sometimes driven by negative experiences. Circuit breaker types are not equal with respect to TRVs; for example, air-blast and SF_6 gas types are sensitive to the rate of rise of the TRV (RRRV), but bulk-oil and vacuum types tend to have a low sensitivity to the RRRV because of the effects of post-arc current. For this reason, the short-line fault test did not become relevant until the first applications of air-blast circuit breakers.

The Bibliography chronologically traces the evolution of TRVs to what they are today in the IEC and IEEE circuit breaker standards, now mostly harmonized. The literature tracks AIEE- and IEEE-published papers to cover the understanding and development of TRVs in North America directed toward IEEE circuit breaker standards and CIGRE reports for the rest of the world directed toward the corresponding IEC standards.

TRVs involve AC recovery voltages, oscillatory circuits and travelling waves, the basic theory for which was developed by engineers at General Electric (GE), Westinghouse and the American Telephone and Telegraph Company (AT&T) between 1918 and 1933. AC recovery voltages and associated pole factors are calculated using the symmetrical component method discovered by Fortescue of Westinghouse [3] and expanded on by Wagner and Evans of Westinghouse [19] and Clarke of GE [41]. Oscillatory circuits were investigated by Carson [4,9] and Johnson [10], both of AT&T. Carson used the Heaviside operator approach for the calculations. Travelling wave theory was developed by Bewley of GE [49]. Later noteworthy work on pole factors in particular is that by Hammarlund of ASEA [38].

Theory aside, the development of circuit breakers progressed for many years without any realization of the existence of TRVs. Circuit breaker ratings were discussed in the very earliest AIEE meetings [1,2]. No short-circuit test laboratories existed yet, and current interruption tests were performed in the field [5–8,11,16,17,21,23]. Some of these tests were performed by the customer utilities [11]. Interruption was treated as a pure power frequency event based on short-circuit MVA, and the first reference to a further quantity of possible interest appears to be that by Hilliard of GE in 1927 in the discussion to Ref. [11]. Hilliard wrote that interruption is dependent on the "voltage kick" after each current zero and the rate at which this "transient voltage is built up." Hilliard's observation represents the first step toward the understanding and appreciation of transient recovery voltages.

TRV evolution is discussed in the following sections by the type of current to be interrupted, measurements and calculations and finally representation in standards.

H.2 TRVs: Terminal Faults

The earliest AIEE literature reflects an ongoing rivalry between GE and Westinghouse, the principal players being Prince and Skeats at GE and Slepian and Van Sickle at Westinghouse. The literature is reviewed in chronological order in Table H.1.

In summary, starting from Hilliard's observation in 1927, the first TRV standards were published in the 1950s (discussed in the next section). Before this, the TRV values used for oil circuit breaker testing were *de facto* standards, but not always viewed positively by the user utilities. Actual system characteristics were eventually recognized with emphasis primarily placed on the RRRV in North America and the system natural frequency in Europe and elsewhere. The first standardized TRVs were single frequency and evolved into multiple frequency at higher fault current levels after the influence of travelling waves was considered. The IEC and IEEE standards for TRVs developed independently but are now harmonized.

Appendix H: Evolution of Transient Recovery Voltages

Table H.1 Historical development of terminal fault TRVs.

Reference no., year	Author(s)	Comment
[12], 1928 [13], 1929	J. Slepian	Studied short and long arcs and concluded that interruption is dependent on speed with which contact gap recovers after current zero; recovery shown to be initial jump followed by gradual increase in dielectric strength controlled by rate of deionization of contact gap; became basis for "race theory," race between RRDS and RRRV but does not take remnant thermal effects of the arc into account (arc models followed much later in 1939 [32] and 1943 [37]); TRV—referred to as the "re-establishing voltage"—shown first as oscillatory and then exponential by addition of resistor across the circuit breaker terminals, possible first use of an opening resistor (see Section 4.4).
[14], 1930	J. Slepian	Deion circuit breaker described uses air as the interrupting medium; arc drawn between arcing contacts and blown magnetically between copper plates into a series of shorter arcs supposedly beneficial to interruption; reaction of users to non-oil circuit breakers is positive; clearly, oil circuit breakers were viewed as a hazard and Dean (of unknown affiliation) referred to them as "infernal things" in the discussion to the paper.
[17], 1931	R.M. Spurck H.E. Strang	Describes testing of two GE oil circuit breakers on power system with fault levels up to 7 kA, RRRV varied by switching transmission lines in and out, giving 0.27 kV/μs (three lines connected), 0.6 kV/μs (one line connected) and 2.4 kV/μs (no lines connected); test traces show overdamped TRV with travelling wave reflections for first case and underdamped oscillation for the last case; circuit breakers exhibited a significant increase in arcing time with decreasing fault current level.
[20], 1933	R.C. Van Sickle W.E. Berkey	Insightful and significant paper; key details are as follows: • Existence of heretofore unrecognized interaction between arc and circuit. • Period of significance: 100 μs before and 100 μs after current interruption with first few microseconds of particular impact, that is, post-arc current. • RRRV inherently dependent on circuit but influenced by post-arc current producing a resistive damping effect. • Charging of local capacitance before current zero caused by changing arc voltage diverting current from arc.

(*continued*)

Table H.1 (*Continued*)

Reference no., year	Author(s)	Comment
		Discussion equally interesting with critical reactions from Prince and Skeats of GE:
		• Prince rejects the results as being influenced by the inadequacy of the circuit breaker described as "plain break oil circuit breakers notoriously erratic in performance," later as "fair weather circuit breakers" and finally "oldest known form of circuit breakers."
		• On the positive side, Prince proposes that some minimum recovery voltage rates should be established in standards.
		Skeats argues against using RRRV as a basis for type testing and suggests that a "system characteristic be defined as the curve which would be obtained with an ideal circuit breaker," perhaps the first intimation of the inherent TRV approach used in standards.
[21], 1933	R.M. Spurck W.F. Skeats	Circuit breaker testing is discussed, including positive effects of added capacitance and negative effect of series reactors; proposes extrapolation of laboratory test results to more onerous field requirements on short-circuit MVA basis stating that it can be done "with confidence"; in the discussion, users are less than enthusiastic with one (Sporn) asking why the authors did not state that the extrapolation could be done "with safety."
[26], 1938	R.D. Evans A.C. Monteith	Study in which consideration is given to TRV peak values and time to peak in addition to RRRV; discussion by Prince and Van Sickle, among others, shows that overall understanding of TRVs is still illusive.
[28], 1939	H.A. Peterson	Similar study to that in Ref. [26] with more detail; X_0/X_1 ratio reported for first time and angle of fault initiation varied, but no mention made of asymmetry; overshoots—not yet referred to as amplitude factors—of 1.5–1.75 pu are noted.
[29], 1939	C. Concordia W.F. Skeats	Perhaps a first consideration of voltage escalation caused by repeated re-striking during fault clearing and capacitive current switching.
[30], 1939	R.D. Evans A.C. Monteith R.L. Witzke	Study performed on a "calculating board"; switching overvoltage magnitudes are addressed rather than TRVs as such.
[31], 1939	R.C. Van Sickle	Significant for use of oscillating circuit calculations and for showing value of closing resistors to limit capacitor bank inrush currents and opening resistors to limit TRVs.
[33], 1942	H.P. St. Clair J.A. Adams	Reports on survey by utility organization (Association of Edison Illuminating Companies) of power system TRVs by calculation; apparent first use of term transient recovery voltage and showing TRVs as an envelope.

Table H.1 (*Continued*)

Reference no., year	Author(s)	Comment
[34], 1942	R.C. Van Sickle	State-of-the-art paper with respect to TRVs and circuit breaker types; TRVs shown to range from underdamped oscillations to overdamped oscillations when transmission lines are involved; all circuit breaker types of the time exhibited post-arc current and consequent damping of the TRVs.
[38], 1946	P.E. Hammarlund	Landmark document (actually a report) for a number of reasons: • Pole factor calculation and probably the most referenced document in this regard (see Section H.6). • TRV calculation for hundreds of circuit breakers in Swedish power systems 6–200 kV based on Boehne's approach [22]. • Apparent first-time use of current injection device to measure TRVs on a power system (see also next citation).
[39], 1948	P.E. Hammarlund O. Johansen	TRVs calculations with RRRV chosen as key attribute and related to short-circuit MVA showing a decrease in RRRV with increasing MVA; maximum value of RRRV 6.6 kV/μs compared with 10 kV/μs in Ref. [33]; can be argued that the paper represented a shift in TRV standardization from North America to Europe, principally within CIGRE.
[45], 1950	J.S. Cliff	Paper from short-circuit testing laboratory perspective; uses term re-striking voltage for TRVs, proposes definitions for AC recovery voltages, inherent TRVs, RRRVs and (for an apparent first time) amplitude factors; an empirical approach for multiple-frequency TRVs is also proposed.
[46], 1950	L. Gosland J.S. Vosper	TRVs calculated for 66 kV systems in the United Kingdom with RRRV taken as the attribute of significance; results compare well with those of Hammarlund [38,39]; maximum RRRV value calculated was 2.6 kV/μs.
[48], 1951	C.H.W. Lackey	Significant but not-well-known paper describes state of the art for circuit breakers with respect to fault clearing, out-of-phase switching and inductive and capacitive load switching; effect of resistors on TRV damping uses a simple parallel oscillatory circuit (see Figure 4.10) giving the method for TRV calculation described in Section 2.5 and applied in Chapter 4.
[54], 1952	L. Gosland J.S. Vosper	TRVs calculated for 132 kV systems in the United Kingdom; RRRVs decrease with increasing short-circuit MVA with most 5 kV/μs or less and amplitude factors in range 1.2–1.8 pu.
[55], 1953	E.B. Rietz J.W. Beatty	Paper not of great interest, but discussion by European circuit breaker manufacturers is of value; Flurscheim comments that TRVs are related to the parts of the system contributing fault current (higher values through

(*continued*)

Table H.1 (*Continued*)

Reference no., year	Author(s)	Comment
		interconnections at the circuit breaker voltage and lower values through transformers only) with a quasi-standard of 4 kV/μs at rated short-circuit MVA providing adequate safety margins; Meyer notes proposals to correlate natural system frequencies to system voltages at circuit breaker ratings.
[56], 1953	I.B. Johnson A.J. Schultz W.F. Skeats	TRV studies up to 330 kV with calculated RRRV values ranging from 1 to 2 kV/μs and up to 7 kV/μs for the lowest fault current levels.
[57], 1954	AIEE Committee	Report recommended adoption of a symmetrical current rating basis to replace the total current MVA basis.
[62], 1956	S. Fukuda F. Mori	TRV standardization in Japan is proposed on the basis of system natural frequency (instead of related RRRV value) and amplitude factor.
[66], 1958	S. Fukuda F. Mori K. Nakanishi J. Ushio	Further consideration of TRVs in Japan with particular reference to AC recovery voltage contributions, that is, pole factors.
[67], 1958	A. Hochrainer	Proposes a method to deal with multiple-frequency TRVs in test circuits; an equivalent four-parameter method is described and eventually incorporated into the IEC circuit breaker standard in 1971.
[69], 1960	H. Meyer	Discusses issues of the time, including four-parameter TRV testing, short-line faults, evolving faults, capacitor bank switching and synthetic testing.
[70], 1960	P. Baltensperger	Test circuits to achieve four-parameter TRV waveshapes are described.
[75], 1962	H. Meyer	Proposes specification for four-parameter TRVs (still referred to as re-striking voltages) including RRRV and amplitude factor; four parameters are designated as e_1, t_1, e_c and t_c, which evolved into the e_1, t_1, u_c and t_2 parameters of today, respectively.
[76], 1962	P. Baltensperger	This paper is probably the first time that the concept of a standardized circuit for TRV determination is proposed. Refer to Chapter 4.
[77], 1962	J. Zaborszky J.W. Rittenhouse	Graphical calculation of the AC recovery voltages associated with switching resistive, inductive and capacitive loads is described.
[78], 1964	P. Baltensperger	Apparent first time use in a CIGRE and IEC context of term "transient recovery voltage" replacing "re-striking voltage."
[80], 1965	O. Naef C.P. Zimmerman J.E. Beehler	Classification by utility authors that greatly influenced the development of TRV standards in North America; discussion by Soles describes the exponential–cosine

Appendix H: Evolution of Transient Recovery Voltages 217

Table H.1 (*Continued*)

Reference no., year	Author(s)	Comment
		TRV later incorporated into the IEEE circuit breaker standard.
[81], 1966	A. Eidinger A. Mosbeck	This report makes the case to treat RRRV as a true rate of rise and no doubt represents a first move toward defining the RRRV as a tangent to the TRV initial waveshapes.
[82], 1966	P. Baltensperger	TRV studies on the 245 kV Italian system gave amplitude factors of 1.4 pu at the highest fault current levels and close to 2 pu for the lowest (see also next citation).
[84], 1967	G. Catenacci L. Paris J.-P. Couvreux M. Pouard	Studies on 245 kV Italian system with a maximum fault level of 33 kA: • Fault fed by single transformer, single-frequency TRV with amplitude factor of 1.8 pu. • Fault currents up to 50% with one line connected, four-parameter TRV. • Fault current between 50 and 100% with two or more lines connected, also a four-parameter TRV.
[85], 1968	A. Colombo *et al.*	Companion paper to the previous citation describing the system representations used.
[86], 1968	J. Buter E. Markworth R. Richter	Describes terminal fault tests on a 220 kV system; measured TRVs were within the parameters under discussion at the time.
[87], 1968	P. Baltensperger *et al.*	Review of all the TRV studies initiated through CIGRE; using a four-parameter TRV for all but the transformer-fed fault is proposed and (for the first time) a fifth parameter, the time delay, is added; the RRRV for the highest fault level was concluded to be less than 1 kV/µs.
[88], 1968	E. Ruoss	Test circuits for fault current and load current switching are described.
[90], 1969	R.G. Colclasser D.E. Buettner	This is a much cited paper that influenced the development of TRV standards and is applicable to the T100 and T60 test duty cases; the discussion to the paper by Beehler is of great interest, showing the results of an actual fault test clearly demonstrating the effect of lines (and no lines) on TRVs.
[95], 1984 [97], 1988	C.L. Wagner H.M. Smith S.S. Berneryd	These two papers analyze the IEC and IEEE approaches to TRV standardization; a strong case is made for harmonizing to the former approach, which has now occurred; delayed by some 20 years, this topic is discussed further in the next section.

H.3 Terminal Fault TRV Standardization

Circuit breaker standardization within IEC started in the 1920s, and the first edition of IEC 56 (the forerunner of IEC 62271-100) appeared in 1937. This edition did not address TRVs. Three further editions of IEC 56 followed:

- *Second edition, 1954*: Short-circuit ratings expressed in MVA. TRV single frequency and referred to as a re-striking voltage; first-pole-to-clear factor 1.5 pu; five test duties, test duties 1–4 being 10, 30, 60 and 100% symmetrical and test duty 5 being 100% asymmetrical.
- *Third edition, 1971*: Symmetrical current based using the R10 series; two- and four-parameter TRVs (now referred to as such), and first-pole-to-clear factors of 1.3 and 1.5 pu; five test duties as for the second edition with rates of rise of 7, 5, 2 and 1 kV/μs for test duties 1–4, respectively.
- *Fourth edition, 1987*: Changes compared to the third edition include relating the first-pole-to-clear factors to system voltage as it is today and the rates of rise of 7, 5, 3 and 2 kV/μs for test duties 1–4, respectively.

The fourth edition of IEC 56 was revised in 2001 to become the IEC 62271-100, Edition 1. The test duties became T100s, T100a, T60, T30 and T10 as discussed in Chapter 4.

Standardization within North America through IEEE followed a similar pattern. Single-frequency TRVs were adopted at first for all short-circuit levels and later refined to a so-called exponential–cosine TRV at 100 and 60% short-circuit current levels. This type of TRV might well be described as a three-parameter TRV (two voltage parameters and one time parameter), where the initial part is an exponential function and the second part a damped single-frequency oscillation as shown in Figure H.1. Note that the two parts are not additive but occur in sequence. The values for E_1 and E_2 are given by

$$E_1 = 1.5 \frac{\sqrt{2}}{\sqrt{3}} U_r = 1.225 U_r$$

and

$$E_2 = 1.76 U_r,$$

with 1.5 being the k_{pp} value used for all voltage ratings. T_2 and R values are given by voltage rating.

Taking the case of a 245 kV circuit breaker, $U_r = 245$ kV, $T_2 = 520$ μs and $R = 1.8$ kV/μs (taken from IEEE C37.06).

$$E_1 = 300 \text{ kV},$$
$$E_2 = 431 \text{ kV}.$$

The part of the TRV is exponential given by the equation

$$v_e = 300 \left(1 - e^{-t/\tau}\right)$$

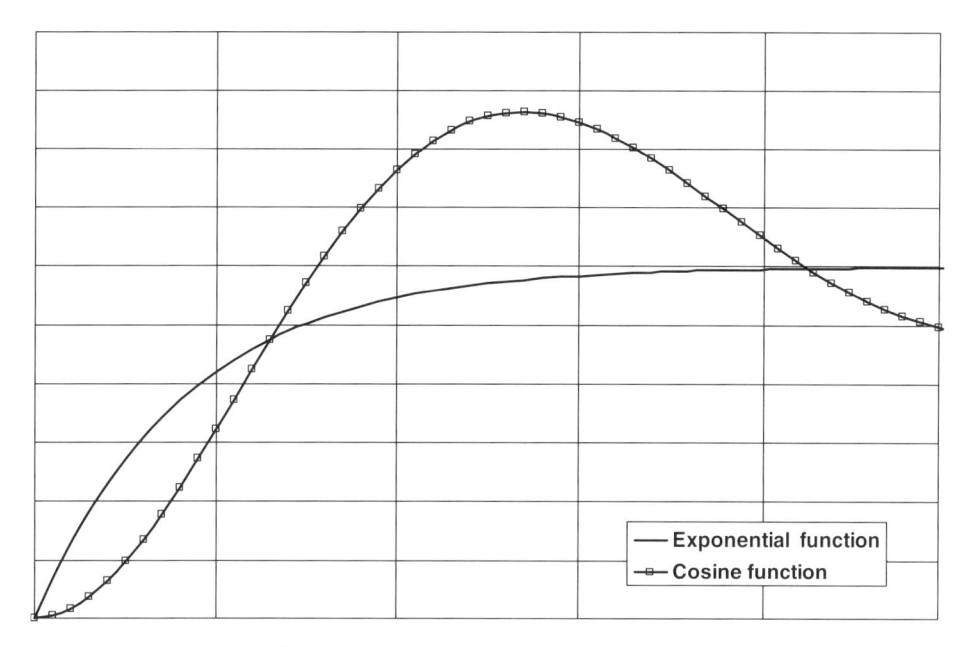

Appendix H: Evolution of Transient Recovery Voltages

Figure H.1 IEEE exponential–cosine TRV representation.

and the time constant is determined from R.

$$\frac{dv}{dt} = \frac{300}{\tau} e^{-t/\tau}$$
$$= \frac{300}{\tau} \quad \text{at } t = 0$$
$$= 1.8,$$
$$\tau = 166 \, \mu s,$$

and the required equation is

$$v_e = 300(1 - e^{-6024t}), \tag{H.1}$$

where t is in seconds.

For the oscillatory part, the degree of damping d_p and the per-unit values of t_g need to be determined. E_2 can be rewritten as

$$E_2 = 1.225 k_{af} U_r,$$

giving $k_{af} = 1.44$.

From Eq. (2.46), $d_p = R_C/R = 0.253$.

$$1 \text{ pu } t_g = T_2/\pi$$
$$= 165.5 \, \mu s.$$

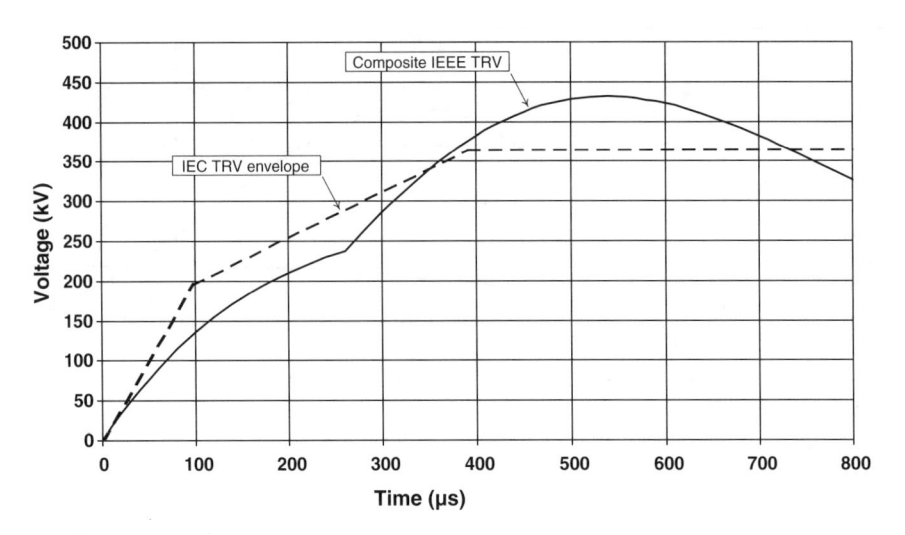

Figure H.2 IEEE exponential–cosine TRV for a 245 kV rated circuit breaker.

The equation for the oscillatory part is

$$v_o = 300\left[1 - e^{-0.253 t_g}\left(\cos 0.967 t_g + 0.26 \sin 0.967 t_g\right)\right]. \tag{H.2}$$

Equations (H.1) and (H.2) are plotted in Figure H.2. From 0 to 260 μs, the TRV is exponential and from 260 μs onward the TRV is oscillatory. The TRV envelope for 245 kV according to IEC is also shown. The differences in the peak values are due to the lower pole factor k_{pp} for the IEC case.

H.4 Short-Line Fault

At the beginning of the 1950s, the short-line fault was not recognized as being an issue. State-of-the-art papers by Meyer [44] in 1950 and Lackey [48] in 1951 make no mention of it at all. This is most probably because of the predominant use of oil circuit breakers, which are not sensitive to RRRV. With the increasingly widespread use of air-blast circuit breakers, which are sensitive to RRRV, field experience showed that the short-line fault was a significant issue. In fact, experience showed that a fault at a specific distance along the line was most onerous for air-blast circuit breakers, as is also the case for the later SF_6 circuit breakers.

Descriptions of the duty, impact on circuit breakers and laboratory testing can be found in Refs [64,68,73–79]. Colclasser et al. showed that the fault duty can be made less onerous by the addition of capacitance on the line side of the circuit breaker [92]. This remedy was applied in North America for a number of years, particularly on SF_6 dead-tank type breakers.

Standardization occurred much more rapidly than for terminal faults. Based on a proposal by Catenacci in 1964 (see Appendix II of Ref. [78]), the short-line fault was incorporated into the third edition of IEC 56 in 1971. Various line surge impedances were allowed dependent on the conductor configuration; however, in the fourth edition in 1987, the surge impedance was fixed at 450 Ω.

H.5 Inductive and Capacitive Load Current Switching

Interrupting inductive and capacitive load currents were recognized early on as a difficult switching duty mostly with respect to the overvoltages produced. This applied to both oil and air-blast circuit breakers and the most common remedial measure was to incorporate opening resistors [47,48]. Oil circuit breakers presented an added difficulty, that is, being self-blast, the low current values associated with these switching duties are not conducive to early interruption.

Inductive load currents were originally described as small inductive currents, a term that carried into the 1980s. Early papers by Baltensperger addressed the duty noting the occurrence of current chopping [44,47]. The first field test on a high-voltage shunt reactor was performed in 1951 and is described in Ref. [53]. Further papers followed without treating the interactive nature of the duty [71,76,77]. The phenomenon of current chopping was not understood until the work of Rizk in the 1960s (a discussion of Rizk's work can be found in Ref. [102]). This enabled accurate calculations for the duty and standardized approach to the application.

With respect to standardization, the subject was discussed in successive meeting within CIGRE [78,82], and a long-running working group was established in the late 1970s. The working group published a number of documents collected in Ref. [101]. This work led to an IEC Technical Report in 1994, which covered both type testing and an application guide [100]. In parallel with this, IEEE produced an application guide (IEEE C37.015), and some utilities resorted to conducting field tests to verify circuit breaker capability [86,94,96,98,99]. IEC Technical Report 1233 was later split between application [102] and type testing [103].

Capacitive load currents were easier to deal with after the basic principles were established [52,58]. The cases of single capacitor banks, lines and cables were introduced into both the IEC and IEEE standards in 1971 and have evolved to the C1 and C2 approach of today.

H.6 Terminal Fault TRV Calculation

TRVs, as we have learned, comprise two components: the AC component as determined by pole factors and the oscillatory component as determined by the applicable RLC circuit. The evolution of the calculation of each will be considered separately.[1]

H.6.1 Pole Factor Calculation

An early, if not the first, consideration of AC recovery voltages is that of Park and Skeats [15]. The method used is to apply superposition symmetrical component theory to derive the general pole factor equations for the first- and second-pole-to-clear for a three-phase-to-earth fault in an earthed neutral system. The method determines the effective impedance seen by the first- and second-pole-to-clear for a unit current. The recovery voltage in turn is the injected (and cancelling) fault current times the effective impedance according to superposition theory.

The circuit for the first-pole-to-clear is shown in Figure H.3. The convention is the same as that used in Chapter 3, that is, the fault is fed by a balanced system to the left but not shown. The

---

[1] The calculation of fault current levels is outside this scope, and readers are referred to Refs [19,51,59,61] in this regard.

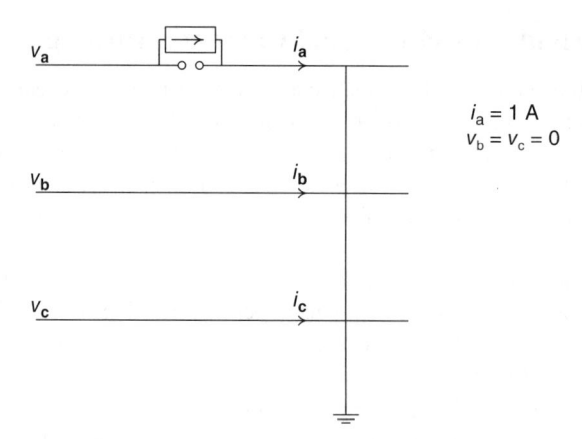

Figure H.3 Circuit for first-pole-to-clear pole factor calculation [15].

calculation (similar to those in Chapter 3 and not repeated here) shows that the effective impedance X_{e1} is given by

$$X_{e1} = \frac{3X_1 X_2 X_0}{X_1 X_2 + X_1 X_0 + X_2 X_0}.$$

From Section 3.1, the fault current I_a is given by

$$I_a = \frac{V_{af}}{X_1}$$

and

$$v_a = X_{e1} I_a$$
$$= V_{af} = \frac{3 X_1 X_0}{X_1 X_2 + X_1 X_0 + X_2 X_0}.$$

The pole factor k_{pp1} is then given by

$$k_{pp1} = \frac{v_a}{V_{af}}$$
$$= \frac{3 X_1 X_0}{X_1^2 + 2 X_1 X_0} \quad \text{for } X_1 = X_2$$
$$= \frac{3(X_0/X_1)}{1 + 2(X_0/X_1)}.$$

For the second-pole-to-clear, the circuit is as shown in Figure H.4. The calculation shows that the effective impedance X_{e2} is given by

$$X_{e2} = \frac{X_1 X_2 + X_1 X_0 + X_2 X_0}{X_1 + X_2 + X_0}.$$

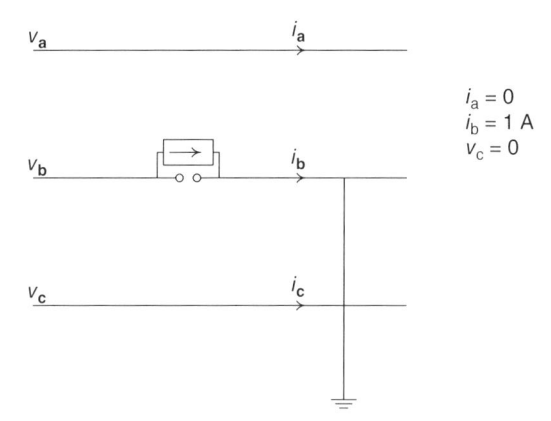

Appendix H: Evolution of Transient Recovery Voltages

Figure H.4 Circuit for second-pole-to-clear pole factor calculation [15].

The fault current I_b follows from the calculation in Section 3.2 and is given by

$$I_b = V_{af} \frac{(a-a^2)X_0 + (a-1)X_2}{X_1X_2 + X_1X_0 + X_2X_0}.$$

Taking the absolute value,

$$|I_b| = \sqrt{3}V_{af} \frac{\sqrt{X_0^2 + X_0X_2 + X_2^2}}{XX_2 + X_1X_0 + X_2X_0},$$

$$v_b = X_{e2}|I_b|$$

$$= \sqrt{3}V_{af} \frac{\sqrt{X_0^2 + X_0X_2 + X_2^2}}{X_1 + X_2 + X_0}$$

and

$$k_{pp2} = \frac{V_b}{V_{af}}$$

$$= \frac{\sqrt{3}\sqrt{X_0^2 + X_0X_1 + X_1^2}}{X_0 + 2X_1} \quad \text{for } X_1 = X_2.$$

Although not covered in the reference, for the third-pole-to-clear,

$$X_{e3} = \frac{1}{3}(X_1 + X_2 + X_0),$$

$$I_c = \frac{3V_{af}}{X_1 + X_2 + X_0},$$

$$v_c = V_{af} \quad \text{and} \quad k_{pp3} = 1.$$

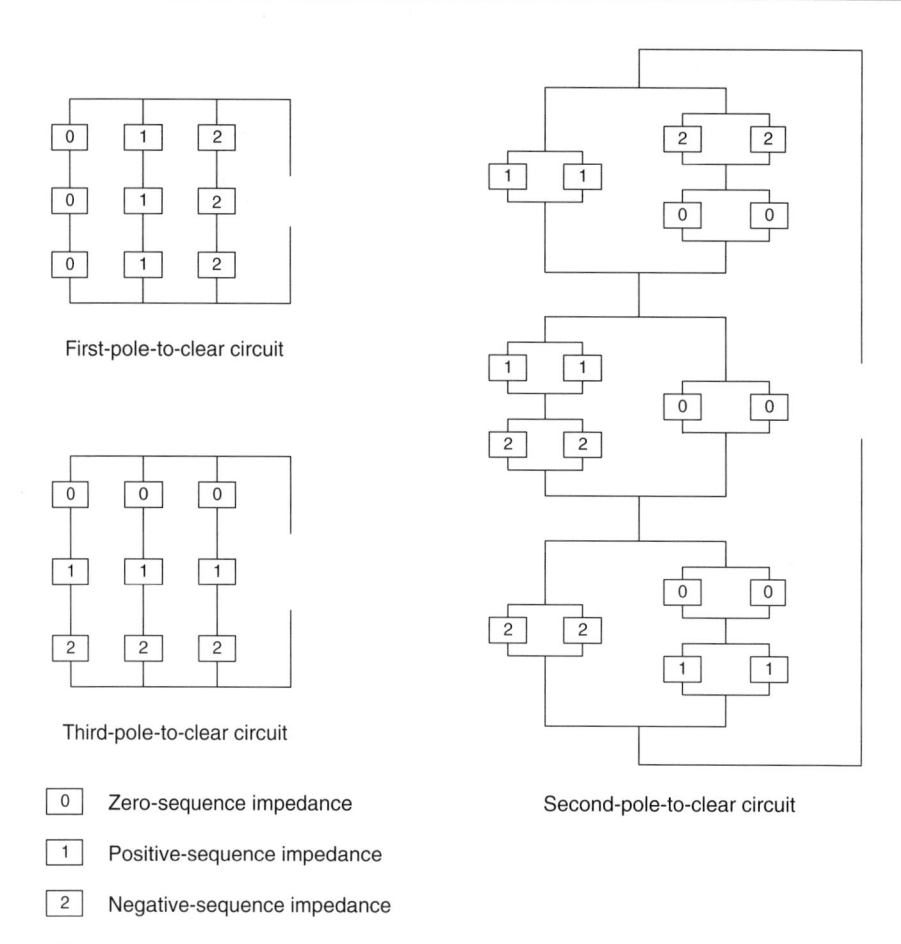

Figure H.5 Impedance circuits for first-, second- and third-pole-to-clear pole factor calculation [38].

Wagner and Evans provide calculations of the first- and second-pole recovery voltages during the interruption of a three-phase-to-earth fault ([19], pp. 44–47). However, the voltages are expressed in symmetrical component terms, and no attempt is made to determine the absolute values and corresponding pole factors.

The most cited reference with respect to pole factors is that of Hammarlund [38]. The reference is of more interest for its two proposals with respect to recovery voltage calculation rather than actual calculation. The first proposal is that the pole factors can be calculated by inserting an impedance across the pole in question, calculating the phase voltage and then allowing the impedance to approach infinity. No attempt is made to verify the proposal, which is correct for the first-pole-to-clear (refer to Section 3.2) and the third-pole-to-clear, but doubtful for the second-pole-to-clear. In the second proposal, impedance circuits are given for which, when unit current is injected, the consequent voltages are the pole factors. The impedances are the fault impedances as calculated by Park and Skeats [15], and the circuits are shown in Figure H.5. No actual calculation is given, nor is any reference made to Ref. [15]

Appendix H: Evolution of Transient Recovery Voltages 225

in this context. The circuit for the second-pole-to-clear is especially complex and perhaps is reverse calculated from the impedance equations.

In summary, the first published calculation of pole factors appears to be that of Park and Skeats [15]. Wagner and Evans provided the first textbook treatment of symmetrical components [19]. The much cited reference by Hammarlund has more to do with possible methods to calculate pole factors rather than actual calculations [38].

H.6.2 Transient Calculation

The literature in this regard is reviewed in chronological order in Table H.2. In summary, key considerations in the development and understanding of current interruption transients are the following:

- The use of the oscillatory circuit theory from the start [18].
- The use of a generic approach based on Ref. [48].
- The fact that approximate calculation methods will yield quite accurate results to complement the complex simulation methods available today [65].

Table H.2 Development of current interruption transient calculation.

Reference no., year	Author(s)	Comment
[18], 1931	S.S. Attwood W.G. Dow W. Krausnick	Even though arcs in air were studied, the study was intended to provide an understanding of the effects of parallel resistance and capacitance on arcs in oil circuit breakers. For the transient calculations, oscillatory circuit analysis is used, a parallel oscillatory circuit for the resistance case and a series oscillatory circuit for the capacitor case. The paper reflects the approach taken in this textbook and by others for the calculation of current interruption transients.
[22], 1935	E.W. Boehne	The paper describes the notion of using cancelling current injection into a circuit to determine its transient response. Although probably the first published paper on the subject, Bewley suggests that the original idea is due to Heaviside [27]. Johnson's conversion method issued to calculate the transient response of coupled circuits [10] and its accuracy and limitations are discussed in the main text.
[24], 1937	R.D. Evans A.C. Monteith	Of historical interest only, the paper describes the use of a calculating board (perhaps a forerunner of transient network analyzers) to calculate TRVs.
[25], 1937	W. Wagner J.K. Brown	Capacitances associated with generators and transformers are distributed rather than lumped elements. Circuit reduction calculations are done to derive equivalent lumped capacitance values for comparison with values derived by actual short-circuit testing.

(continued)

Table H.2 (*Continued*)

Reference no., year	Author(s)	Comment
[27], 1939	L.V. Bemley	Original description of travelling wave phenomena related to switching and used in the derivation of the standard TRVs for T100 and T60 test duties (refer to Ref. [90]).
[35], 1942	J.A. Adams W.F. Skeats R.C. Van Sickle T.G.A. Sillers	The paper is a joint effort by a utility engineer and manufacturing rival engineers from Allis-Chalmers, GE and Westinghouse. The calculation method is based on Johnson [10] and Boehne [22], and the results have no real significance.
[36], 1943	R.D. Evans R.L. Witzke	The "practical" method of transients described is far removed from being understandable or useable.
[38], 1946	P.E. Hammarlund	TRV calculations are based on Johnson [10] and Boehne [22] and have limited value (the calculations are also described in Ref. [39]).
[48], 1951	C.H.W. Lackey	The author appears to be the originator of the idea to treat transients on a generic basis. The calculations in Chapter 2 are based on this approach.
[60], 1955	W.C. Kotheimer	The paper describes the use of a half-wave rectified current pulse to determine circuit inherent TRVs, a practice used today.
[65], 1958	S.B. Griscom D.M. Sauter H.M. Ellis	Approximate calculations for a large number of fault cases are compared with detailed calculations. The former are found to be quite accurate, and the methodology described is very educational.
[91], 1971	A. Greenwood	A well-known textbook with one chapter on oscillatory circuits. The Laplace transform approach is used to derive the generic solutions as originated by Lackey [48].
[93], 1972	E. Slamecka W. Waterschek	German textbook covering a very wide range of circuits associated with current interruption and their transient response using Laplace transforms.

Bibliography

1. Hewlett, E.M. (1916) Rating of oil circuit breakers. *AIEE Trans.*, **35**, 1531–1549.
2. Hewlett, E.M., Mahony, J.N. and Burnham, G.A. (1918) Rating and selection of oil circuit breakers. *AIEE Trans.*, **37**, 123–165.
3. Fortescue, C.L. (1918) Method of symmetrical co-ordinates applied to solution of polyphase networks. *AIEE Trans.*, **37**, 1027–1140.
4. Carson, J.R. (1919) Theory of the transient oscillations of electrical networks and transmission systems. *AIEE Trans.*, **38**, 345–427.
5. Louis, H.C. and Bang, A.F. (1922) Baltimore oil circuit breaker tests. *AIEE Trans.*, **41**, 640–646.
6. Hilliard, J.D. (1922) Tests on general electric oil circuit breakers at Baltimore. *AIEE Trans.*, **41**, 647–652.
7. MacNeill, J.B. (1922) Tests on Westinghouse oil circuit breakers at Baltimore. *AIEE Trans.*, **41**, 653–669 (includes discussion on this paper and the above two papers).
8. Hilliard, J.D. (1924) Oil circuit breaker investigations as carried on with a 26,700 Kv-a generator. *AIEE Trans.*, **43**, 641–659.
9. Carson, J.R. (1926) *Electric Circuit Theory and the Operational Calculus*, McGraw-Hill.

Appendix H: Evolution of Transient Recovery Voltages

10. Johnson, K.S. (1927) *Transmission Circuits and Telephonic Communication*, D. Van Nostrand Company, Inc. (third printing; first printing in 1924).
11. Sporn, P. and St. Clair, H.P. (1927) Tests on high- and low-voltage oil circuit breakers. *AIEE Trans.*, **46**, 289–314.
12. Slepian, J. (1928) Extinction of an A-C arc. *AIEE Trans.*, **47**, 1398–1407.
13. Slepian, J. (1929) Theory of the Deion circuit-breaker. *AIEE Trans.*, **48**, 523–527.
14. Slepian, J. (1930) Extinction of a long A-C arc. *AIEE Trans.*, **49**, 421–430.
15. Park, R.H. and Skeats, W.F. (1931) Circuit breaker recovery voltages – magnitudes and rates of rise. *AIEE Trans.*, **50**, 204–238.
16. Prince, D.C. and Skeats, W.F. (1931) The oil-blast circuit breaker. *AIEE Trans.*, **50**, 506–512.
17. Spurck, R.M. and Strang, H.E. (1931) Circuit breaker field tests on standard and oil-blast explosion-chamber oil circuit breakers. *AIEE Trans.*, **50**, 513–521.
18. Attwood, S.S., Dow, W.G. and Krausnick, W. (1931) Re-ignition of metallic A-C arcs in air. *AIEE Trans.*, **50**, 854–868.
19. Wagner, C.F. and Evans, R.D. (1933) *Symmetrical Components*, McGraw-Hill.
20. Van Sickle, R.C. and Berkey, W.E. (1933) Arc extinction phenomena in high voltage circuit breakers – studied with a cathode ray oscillograph. *AIEE Trans.*, **52**, 850–857.
21. Spurck, R.M. and Skeats, W.F. (1933) Interrupting capacity tests on circuit breakers. *AIEE Trans.*, **52**, 832–840.
22. Boehne, E.W. (1935) The determination of circuit recovery rates. *AIEE Trans.*, **54**, 530–539 (discussion published in *AIEE Trans.*, pp. 191–193, February 1936, and pp. 269–270, March 1936).
23. Prince, D.C. (1935) Circuit breakers for Boulder Dam line. *AIEE Trans.*, **54**, 366–372.
24. Evans, R.D. and Monteith, A.C. (1937) System recovery voltage determination by analytical and A-C calculating board methods. *AIEE Trans.*, **56**, 695–705.
25. Wanger, W. and Brown, J.K. (1937) Calculation of the oscillations of the recovery voltage after rupture of short circuits. *Brown Boveri Rev.*, **24** (11), 283–302.
26. Evans, R.D. and Monteith, A.C. (1938) Recovery-voltage characteristics of typical transmission systems and relation to protector-tubes application. *AIEE Trans.*, **57**, 432–443.
27. Bewley, L.W. (1939) Travelling waves initiated by switching. *AIEE Trans.*, **58**, 18–26.
28. Peterson, H.A. (1939) Power-system recovery-voltage characteristics. *AIEE Trans.*, **58**, 405–413.
29. Concordia, C. and Skeats, W.F. (1939) Effect of re-striking on recovery voltage. *AIEE Trans.*, **58**, 371–376.
30. Evans, R.D., Monteith, A.C. and Witzke, R.L. (1939) Power system transients caused by switching and faults. *AIEE Trans.*, **58**, 386–397.
31. Van Sickle, R.C. (1939) Influence of resistance on switching transients. *AIEE Trans.*, **58**, 397–404.
32. Cassie, A.M. (1939) Arc rupture and circuit severity: a new theory. CIGRE Report 102.
33. St. Clair, H.P. and Adams, J.A. (1942) Transient recovery-voltage characteristics of electric-power systems. *AIEE Trans.*, **61**, 666–669.
34. Van Sickle, R.C. (1942) Transient recovery voltages and circuit breaker performance. *AIEE Trans.*, **61**, 804–813.
35. Adams, J.A., Skeats, W.F., Van Sickle, R.C. and Sillers, T.G.A. (1942) Practical calculation of circuit transient recovery voltages. *AIEE Trans.*, **61**, 771–779.
36. Evans, R.D. and Witzke, R.L. (1943) Practical calculation of electrical transients on power systems. *AIEE Trans.*, **62**, 690–696.
37. Mayr, O. (1943) Beitraege zur Theorie des statischen und dynamischen Lichtbogen. *Electrotechnik*, **37**, 588–608.
38. Hammarlund, P.E. (1946) Transient recovery voltages subsequent to short-circuit interruption with special reference to Swedish power systems. Proceedings No. 189, Royal Swedish Academy of Engineering Sciences.
39. Hammarlund, P.E. and Johansen, O. (1948) Transient recovery voltages subsequent to short-circuit interruption with special reference to Swedish power systems. CIGRE Report 107.
40. Goldman, S. (1949) *Transformation Calculus and Electrical Transients*, Prentice Hall (republished with title *Laplace Transform Theory and Electrical Transients*, Dover Publications, 1955).
41. Clarke, E. (1950) *Circuit Analysis of A-C Power Systems, Volume 1, Symmetrical and Related Components*, John Wiley & Sons, Inc.
42. Rudenberg, R. (1950) *Transient Performance of Electrical Power Systems*, McGraw-Hill (original German version published in 1933).
43. Carter, G.W. (1950) *The Simple Calculation of Electrical Transients*, Cambridge University Press.
44. Meyer, H. (1950) The fundamental problem of high-voltage circuit-breakers. *Brown Boveri Rev.*, **37** (4/5), 108–122.
45. Cliff, J.S. (1950) Testing station re-striking-voltage characteristics and circuit-breaker proving. CIGRE Report 109.
46. Gosland, L. and Vosper, J.S. (1950) Re-striking voltages in British 66kV networks. CIGRE Report 110.

47. Baltensperger, P. (1950) Overvoltages due to the interruption of small inductive currents. CIGRE Report 116.
48. Lackey, C.H.W. (1951) Some technical considerations relating to the design, performance and application of high-voltage switchgear. *Trans. S. Afr. Inst. Electr. Eng.*, 261–298.
49. Bewley, L.V. (1951) *Travelling Waves on Transmission Systems*, Dover Publications (reprinted from original textbook published by John Wiley & Sons, Inc., 1933).
50. Peterson, H.A. (1951) *Transients in Power Systems*, John Wiley & Sons, Inc.
51. Lackey, C.H.W. (1951) *Fault Calculations*, Oliver & Boyd.
52. Van Sickle, R.C. and Zaborszky, J. (1951) Capacitor switching phenomena. *AIEE Trans.*, **70**, 151–159.
53. Baltensperger, P. (1951) Line-dropping and reactor-load interruption tests with a Brown Boveri 220-kV air-blast circuit-breaker in the Stadsforsen power station (Sweden). *Brown Boveri Rev.*, **38** (12), 391–410.
54. Gosland, L. and Vosper, J.S. (1952) Network analyzer and study of inherent re-striking voltage transients on the British 132kV grid. CIGRE Report 120.
55. Rietz, E.B. and Beatty, J.W. (1953) Effect of voltage recovery rates on interrupting performance of air-blast circuit breakers. *AIEE Trans.*, **72**, 303–311.
56. Johnson, I.B., Schultz, A.J. and Skeats, W.F. (1953) System recovery voltage and short-circuit duty for high-voltage circuit breakers. *AIEE Trans.*, **72**, 1339–1347.
57. AIEE Committee (1954) A new basis for rating power circuit breakers. *AIEE Trans.*, **73**, 353–367.
58. Van Sickle, R.C. and Zaborszky, J. (1954) Capacitor-switching phenomena with resistors. *AIEE Trans.*, **73**, 971–977.
59. Lantz, M.W. (1955) Analysis of fault currents for high-voltage circuit-breaker interruption. *AIEE Trans.*, **74**, 41–45.
60. Kotheimer, W.C. (1955) A method for studying circuit transient recovery voltage characteristics of electric power systems. *AIEE Trans.*, **74**, 1083–1087.
61. Skeats, W.F. (1955) Short-circuit currents and circuit breaker recovery voltages associated with 2-phase-to-ground short circuits. *AIEE Trans.*, **74**, 688–693.
62. Fukuda, S. and Mori, F. (1956) Re-striking voltage conditions on the Japanese systems. CIGRE Report 105.
63. Lago, G.V. and Waidelich, D.L. (1958) *Transients in Electric Circuits*, The Ronald Press Company.
64. Skeats, W.F., Titus, C.H. and Wilson, W.R. (1957) Severe rate of rise of recovery voltages associated with transmission line short circuits. *AIEE Trans.*, **76**, 1256–1264.
65. Griscom, S.B., Sauter, D.M. and Ellis, H.M. (1958) Transient recovery voltages on power systems. Part II. Practical methods of determination. *AIEE Trans.*, **77**, 592–604.
66. Fukuda, S., Mori, F., Nakanishi, K. and Ushio, T. (1958) Recovery and re-striking voltages on 66–275kV power systems. CIGRE Report 132.
67. Hochrainer, A. (1958) A proposal for the establishment of an equivalent re-striking voltage for circuit-breakers (four-parameter method). Appendix II. CIGRE Report 151.
68. Petitpierre, R. (1960) Airblast circuit-breakers with relation to stresses which occur in modern networks, with particular reference to the interruption of short-line faults. CIGRE Report 115.
69. Meyer, H. (1960) Report on the activity of CIGRE Study Committee No. 3 (High-Voltage Circuit Breakers). CIGRE Report 138.
70. Baltensperger, P. (1960) Reproduction of the re-striking voltage at the terminals of circuit-breaker characterized by four parameters. Facilities in short-circuit testing stations. Appendix II. CIGRE Report 138.
71. Baltensperger, P. (1960) The shape and magnitude of overvoltages when interrupting small inductive and capacitive currents in H.V. systems. *Brown Boveri Rev.*, **47**(4), 195–224.
72. Baltensperger, P. and Ruoss, E. (1960) The short-line fault in high-voltage systems. *Brown Boveri Rev.*, **47**(5/6), 329–339.
73. Eidinger, A. and Rieder, W. (1962) Short line fault problems. CIGRE Report 103.
74. Frate, G. (1962) Comparison between air-blast and low-oil content circuit-breakers as regards the short-line fault. CIGRE Report 144.
75. Meyer, H. (1962) Report on the activity of CIGRE Study Committee No. 3 (High-Voltage Circuit Breakers). CIGRE Report 148.
76. Baltensperger, P. (1962) New knowledge in the field of switching phenomena and circuit-breaker testing. *Brown Boveri Rev.*, **49**(9/10), 381–397.
77. Zaborszky, J. and Rittenhouse, J.W. (1962) Some fundamental aspects of recovery voltages. *AIEE Trans.*, **81**, 815–821.
78. Baltensperger, P. (1964) Report on the activity of CIGRE Study Committee No. 3 (High-Voltage Circuit Breakers). CIGRE Report 132.

Appendix H: Evolution of Transient Recovery Voltages

79. Eidinger, A. and Jussila, J. (1964) Transients during three-phase short-line faults. *Brown Boveri Rev.*, **51** (5), 303–319.
80. Naef, O., Zimmerman, C.P. and Beehler, J.E. (1965) Proposed transient recovery voltage ratings for power circuit breakers. *IEEE Trans. Power Apparatus Syst.*, **84**, 580–608.
81. Eidinger, A. and Mosbeck, A. (1966) Initial rate of rise of re-striking voltage in three-phase systems. CIGRE Report 129.
82. Baltensperger, P. (1966) Report on the activity of CIGRE Study Committee No. 3 (High-Voltage Circuit Breakers). CIGRE Report 146.
83. Koppl, G. and Geng, P. (1966) Determination of the transient recovery voltages in symmetrical three-phase systems. *Brown Boveri Rev.*, **53** (4/5), 311–325.
84. Catenacci, G., Paris, L., Couvreux, J.-P. and Pouard, M. (1967) Transient recovery voltages in high-voltage networks. *IEEE Trans. Power Apparatus Syst.*, **86**, 1420–1431.
85. Colombo, A., Couvreux, J.-P., Gielssen, W., Kendall, P.G. and Vontobel, J. (1968) Determination of transient recovery voltages by means of transient network analyzers. *IEEE Trans. Power Apparatus Syst.*, **87**, 1371–1380.
86. Buter, J., Markworth, E. and Richter, F. (1968) Field tests in a 220kV network of high short-circuit power. CIGRE Report 13-09.
87. Baltensperger, P., Cassie, A.M., Catenacci, G., Hochrainer, A., Johansen, O.S. and Pouard, M. (1968) Transient recovery voltages in high voltage networks – terminal faults. CIGRE Report 13-10.
88. Ruoss, E. (1968) Test methods and facilities in the high-power testing station. *Brown Boveri Rev.*, **55** (12), 714–726.
89. Rudenberg, R. (1968) *Electric Shock Waves in Power Systems*, Harvard University Press (original German version published in 1923).
90. Colclasser, R.G. and Buettner, D.E. (1969) The travelling wave approach to transient recovery voltages. *IEEE Trans. Power Apparatus Syst.*, **88**, 1028–1035.
91. Greenwood, A. (1971) *Electrical Transients in Power Systems*, John Wiley & Sons, Inc. (*Note*: Chapter on oscillatory circuits based on Greenwood, A.N. and Lee, T.H. (1963) Generalized damping curves and their use in solving power switching transients. *IEEE Trans. Power Apparatus Syst.*, 82, 527–535).
92. Colclasser, R.G., Berkebile, L.E. and Beuttner, D.E. (1971) The effect of capacitors on short-line fault component of transient recovery voltage. *IEEE Trans. Power Apparatus Syst.*, **90**, 660–669.
93. Slamecka, E. and Waterschek, W. (1972) *Schaltvorgange in Hoch- und Niederspannungsnetzen*, Siemens Aktieengesellschaft.
94. Sarkinen, S.H., Schockelt, G.G. and Brunke, J.H. (1979) High frequency switching surges in EHV shunt reactor installation with reduced insulation levels. *IEEE Trans. Power Apparatus Syst.*, **98**, 1013–1021.
95. Wagner, C.L. and Smith, H.M. (1984) Analysis of transient recovery voltage (TRV) rating concepts. *IEEE Trans. Power Apparatus Syst.*, **103**, 3353–3363.
96. Peelo, D.F., Avent, B.L., Drakos, J.E., Guidici, B.C. and Irvine, J.R. (1988) Shunt reactor switching tests in BC Hydro's 500kV system. *IEE Proc.*, **135**, 420–434.
97. Berneryd, S.S. (1988) Improvements possible in testing standards for high-voltage circuit breakers. Harmonization of ANSI and IEC testing. *IEEE Trans. Power Apparatus Syst.*, **3**, 1707–1713.
98. Reckleff, J.G., McCabe, A.K., Mauthe, G. and Ruoss, E. (1988) Application of metal oxide varistors on an 800kV circuit-breaker used shunt reactor switching. CIGRE Report 13-16.
99. McCabe, A.K., Seyrling, G., Mandeville, J.D. and Willieme, J.M. (1991) Design and testing of a three-break 800 kV circuit-breaker with ZNO varistors for shunt reactor switching. Proceedings of the IEEE PES T&D Conference, Dallas, pp. 499–507.
100. IEC (1994) High-voltage alternating current circuit-breakers – inductive load switching. Technical Report 1233, 1st edn.
101. Berneryd, S.S. (1995) Interruption of Small Inductive Currents. CIGRE Technical Brochure 50.
102. IEC62271-306 (2012) *High-Voltage Switchgear and Controlgear: Part 306 – Guide to IEC 62271-100, IEC 62271-1 and Other IEC Standards Related to Alternating Current Circuit-Breakers*, International Electrotechnical Commission.
103. IEC62271-110 (2012) *High-Voltage Switchgear and Controlgear: Part 110 – Inductive Load Switching*, International Electrotechnical Commission.

Index

a-operator, 33–35, 195
Air blast circuit breaker, 83, 211, 212, 220, 221
Air break disconnector, 139, 209
Amplitude factor, 3, 25–27, 61, 64, 85, 92, 99, 128
Arc, 5, 122, 139, 209
    arc voltage, 122–124, 131, 175, 213
    free burning arc, 139
    secondary arc, 135
Arcing, 5, 139, 169
    arcing contacts, 139, 213
    arcing time, 5, 40, 129, 137, 139, 140, 167, 169–174, 209, 213
    arcing window, 167, 170, 172, 173
Asymmetrical current, 2, 93, 95–97, 102, 197
    AC component, 97, 168
    DC component, 95–97, 168
    major loop, 6, 98, 102, 168, 197, 199
    minor loop, 6
    peak value, 95, 96, 98
    time constant, 92, 95–98
Asymmetry, 6, 97, 102, 214

Back-to-back switching, 1, 143, 146–150
Balanced faults, 37, 40,
Balanced source, 53, 54
Break time, 5
Breaking operation
    arcing time, 5, 40, 129, 137, 139, 140, 170, 171, 173, 174
        intermediate, 173
        maximum, 167, 169–173

    minimum, 40, 139, 169, 170, 171, 173
    contact separation, 5, 97, 168, 170
    re-ignition, 1, 2, 137, 143, 156–160
    re-strike (re-striking), 152, 156–160, 165, 166, 207, 209, 218
Bulk oil circuit breaker, 211

Capacitive current switching, 4, 141, 165, 166, 175, 208
    back-to-back switching, 143, 146–150
    cable switching, 163, 164
    inrush current, 8, 141–150, 160, 163–165, 209
    outrush current, 141, 159, 166
    shunt capacitor bank, 73, 83, 102, 141–160, 165
    unloaded transmission line switching, 166
Capacitor bank, 141, 142, 152, 153, 157, 158, 160
    back-to-back capacitor bank, 8, 30, 141, 147,
    earthed neutral, 144, 147, 148, 153, 160, 165
    isolated neutral, 120–122, 130–135
    single capacitor bank, 145, 147, 148, 153, 169, 221
    unearthed neutral, 148, 153, 155, 165, 166
Circuit breaker
    air blast circuit breaker, 83, 211, 212, 220, 221
    arcing contacts, 139, 213
    bulk oil circuit breaker, 211
    closing resistors, 149, 214
    closing time, 5, 149
    contact make, 5

Current Interruption Transients Calculation, First Edition. David F. Peelo.
© 2014 John Wiley & Sons, Ltd. Published 2014 by John Wiley & Sons, Ltd.

Circuit breaker (*continued*)
  contact part, 5, 137, 139
  contact separation, 5, 168
  contact touch, 5, 141, 147, 149
  dual pressure circuit breaker, 211
  interrupting time, 5, 167–173
  main contacts, 139
  minimum oil circuit breaker, 129, 160, 211
  opening resistors, 21, 83, 86, 87, 102, 213, 221
  opening time, 5, 167, 170, 173
  prestrike, 5, 141, 147, 160
  puffer circuit breaker, 211
  SF_6 circuit breaker, 83, 129, 131–135, 139, 211, 220
  single pressure circuit breaker, 211
  vacuum circuit breaker, 4, 133, 134, 139, 140, 160, 173, 175, 211
Controlled switching, 139, 149
Current
  capacitive load current, 141, 167, 221
  capacitive charging current, 169, 207
  chopped current, 125, 128, 129, 140
  chopping current, 122–125, 139, 175, 221
  chopping number, 129, 139, 140
  current injection, 9, 21, 22, 24, 57, 65, 66, 118, 151, 188, 206, 215, 225
  current interruption, 1, 4–6, 40, 47, 53, 66, 91, 98, 104, 122, 124, 127, 139, 142, 150, 152–154, 160, 167, 177, 184, 200, 209, 225
  current limiting reactors, 149
  fault current, 21, 36–42, 61, 71, 72, 84, 86, 92, 99–102, 104, 113, 115, 152, 160, 217
  inductive load current, 32, 167, 221
  inrush current, 142–152, 160, 163–165
  outrush current, 141, 159, 166
  post-arc current, 173, 212, 213, 215
  ramp current, 21–24, 31, 188
  shunt reactor current, 130
  symmetrical current, 168, 216, 218

Damping, 7, 12, 20,
  coefficient of damping, 144, 145
  critically damped, 12, 20, 28, 143, 149, 183
  damping factor, 25, 138, 208
  degree of damping, 8, 12, 19, 23, 24, 25, 27, 30, 31, 72, 144–148, 188, 219
  overdamped, 10, 12, 14, 16–20, 22, 24, 30, 31, 61, 65, 67, 70, 74, 75, 77, 91, 102, 142, 150, 151, 163, 182, 188, 213, 215
  underdamped, 11, 13, 14, 17–20, 22–25, 27, 30, 31, 61, 63, 70, 71, 72, 77, 78, 81, 90, 102, 143, 148, 150, 152, 157, 184, 213, 215
DC component, 95–97, 168
DC current, 95, 96, 179
DC time constant, 95–98, 168, 197
Differential equations, 177–186
  homogeneous, 66, 184,
  non-homogeneous, 8, 21, 68, 184, 185, 186
  undetermined coefficients, method of, 69, 184, 185
Disconnector, 166, 207, 208
  air break disconnector, 120, 139, 209
  GIS disconnector, 209
Duality, principle of, 187–189

Earthing, 3, 59, 120, 130, 139, 160
  effectively earthed, 3, 4, 40, 41, 43–51, 57, 58, 61, 63, 65, 101, 102, 166, 172–173
  impedance earthed, 3, 200
  non-effectively earthed, 3, 4, 40, 41, 51–56, 63, 72, 99, 102, 168–171
  solidly earthed, 3, 169
Extra High Voltage (EHV), 135, 211

Faults
  double earth fault, 99–101, 169
  double line to earth fault, 40, 42, 43
  out-of-phase fault, 3, 4, 90, 91, 102, 167, 169
  short line fault, 3, 4, 104–118, 167, 169, 212, 220
  single phase to earth fault, 40, 101, 118
  terminal fault, 4, 61–103, 212, 217–220
  three-phase to earth fault, 3, 36, 38–41, 63, 151, 171, 221, 224
  three-phase unearthed fault, 3, 41, 42, 168
  transformer-fed fault, 72, 102
First-pole-to-clear factor, 3, 40, 53, 57, 61, 63, 72, 83, 91, 123, 160, 203, 218
First-pole-to-clear representation
  shunt reactor switching, 56, 121, 200–206
  terminal fault, 4, 31, 61–103, 212, 217, 218–220
  unloaded transmission line switching, 160–163
Four-parameter TRV, 71, 99, 216, 216, 217
Free burning arc, 139

Index

Gas Insulated Switchgear (GIS), 166, 207
GIS disconnector, 209

High voltage circuit breakers, 2, 3

IEC 62271-100, 2, 40, 61, 96, 98, 141, 147, 158, 167–170, 173–175, 199, 218
IEC 62271-110, 140, 174
IEC 62271-302, 166
IEC 62271-306, 140, 174
Inductive load switching, 120–140, 174, 175
   shunt reactor switching, 56, 121, 123–130, 139
   unloaded transformer switching, 120, 139, 174
Inherent transient recovery voltage, 173
Inrush current, 142–152, 160, 163–165, 209
Intermediate arcing time, 173
Interrupting time, 5, 167–173, 211

Kilometric fault, 104

Load current switching, 167, 217,
   capacitive current, 167, 221
   inductive current, 167, 221

Major loop, 6, 98, 102, 168, 197
Making operation
   contact touch, 5, 6, 147, 149
   prestrike, 5, 147
Maximum arcing time, 167, 170–173
Minimum arcing time, 40, 139, 169, 170, 172
Minimum oil circuit breaker, 129, 160, 211
Minor loop, 6
Multiple re-strikes, 157

Negative sequence components, 33
Neutral earthing, 120, 130, 139
   effectively earthed, 3, 40, 43–51, 57, 58, 65, 101, 166, 171–173
   neutral impedance earthed, 3, 200
   non-effectively earthed, 3, 4, 40, 41, 51–56, 63, 72, 99, 102, 168–171
   solidly earthed, 3, 169
Neutral shift, 123, 126, 128, 140, 204

Opening resistors, 21, 83, 102, 214, 221
Opening time, 5, 167, 170, 173
Oscillatory circuits, 61, 212, 226
   damping quantities
      coefficient of damping, 144, 145
      damping factor, 25, 138, 208

degree of damping, 8, 12, 19, 23, 24, 25, 27, 30, 31, 72, 144–148, 188, 219
oscillation quantities
   aiming point, 7, 8, 24, 59, 125, 157
   axis of oscillation, 7, 8, 15, 18, 21, 24, 25, 59, 91, 92, 105, 137, 157
   starting point, 8, 15, 18, 24, 25, 105, 151, 157
parallel RLC oscillatory circuits, 8, 9, 18–27, 31, 151, 187, 188
   critically damped, 19–24, 30, 31, 143, 183, 188
   overdamped, 19, 20, 22, 24, 31, 182
   underdamped, 19, 20, 22–25, 31, 184
series RLC oscillatory circuits, 8–18, 30, 125, 139, 151, 188, 207
   critically damped, 11, 13, 14, 16, 18, 30, 31, 183, 188
   overdamped, 10, 12, 14, 16, 17, 30, 182, 188
   underdamped, 11, 13, 14, 17, 18, 30, 139, 157, 188
generic time, 12, 13, 21, 24–26
real time, 13, 75, 91, 143
Out-of-phase switching, 4, 90–92, 102, 169
Outrush current, 141, 159, 166

Pole factors, 33–60, 99, 212, 220, 221–225
   first-pole-to-clear, 5, 40, 42, 45, 51, 53, 56, 57, 160, 200, 203, 218, 222
   second-pole-to-clear, 40, 47, 221, 223
   third-pole-to-clear, 41, 118, 171, 223
Positive sequence components, 33
Post-arc current, 173, 212, 213, 215
Power system earthing, 3
Prestrike, 5, 141, 147, 160
Principle of duality, 187–189
Principle of superposition, 36, 122
Prospective transient recovery voltage (TRV), 71
Puffer circuit breaker, 211

Rate-of-Rise of Dielectric Strength (RRDS), 213
Rate-of Rise of Recovery Voltage (RRRV), 4, 61, 64, 66, 73, 81, 102, 114, 116, 132–137, 173, 212–217
Reactor
   series reactor, 84, 86, 88, 89, 102, 149, 151, 152
   shunt reactor, 120–122, 126, 130–137, 139, 169, 174, 200, 206, 221

Recovery Voltage (RV), 40, 42, 47, 53, 99, 152–156
Re-ignition, 137, 138–140, 156–160
Re-ignition overvoltage, 137, 138, 140
Re-strike, 156–160
Re-striking overvoltage, 152
RLC oscillatory circuits, 8–31

Secondary arc, 135
Self-blast circuit breaker, 221
SF_6 circuit breaker, 129, 132, 133, 135, 220
Short line fault, 3, 4, 104–118, 167, 169, 212, 220
Shunt reactor switching, 121–130, 169, 174, 200–206
  chopping (chopped) current, 122–125, 134, 139, 174, 221
  chopping number, 128, 129, 139, 140
  controlled switching, 139
  first-pole-to-clear factor, 123
  general case, 123–130
  neutral earthing, 120, 130, 139
    directly earthed, 130, 169
    neutral reactor earthed, 135, 136
    isolated neutral, 130–135
  re-ignitions, 143, 156–160
  suppression peak overvoltage, 123–125, 140
Simpson's Rule, 99, 198
Superposition, principle of, 36, 122
Symmetrical components, 33, 36, 40, 42, 99, 225
  a-operator, 33–35
  negative sequence, 33–36, 39, 224
    current, 33, 36
    impedance, 36, 224
    network, 36, 37, 52, 55, 201, 202,
    voltage, 33, 35, 36
  pole factor calculation, 33–60
    double earth fault, 99–101
    terminal faults, 61–102
      first-pole-to-clear, 61, 63–66, 72, 73

    second-pole-to-clear, 41, 51, 54–56, 171, 223
    third-pole-to-clear, 41, 118, 171, 223
  positive sequence, 33, 34, 224
    current, 39
    impedance, 40, 224
    network, 36, 39
    voltage, 36
  zero sequence, 33, 34, 99, 224
    current, 57, 225
    impedance, 224
    network, 36, 39, 44, 48, 52, 57, 202, 225
    voltage, 52, 201, 202

Time constant, 92, 95, 96, 168, 181, 197, 219
Transient Recovery Voltage (TRV), 8, 15, 40, 61, 73, 120, 149, 167, 211–226
  amplitude factor, 60, 61, 70, 75, 99, 214–217
  four parameter TRV, 71, 91, 99, 102, 216, 217
  inherent TRV, 173, 214, 215, 226
  time delay, 4, 61, 62, 64, 71, 74, 76, 85, 217
  two parameter TRV, 71, 72, 218
Travelling waves, 28–32, 67–70, 104, 212
Type testing, 97, 98, 129, 139, 158, 167–175, 214, 221
  short circuit current, 167

Ultra High Voltage (UHV), 83
Undetermined coefficients, method of, 69, 184, 185
Unloaded transformer switching, 139, 174
Unloaded transmission line switching, 160–163, 166

Vacuum circuit breaker, 4, 133, 134, 140, 160, 173, 211
Very Fast Transient Overvoltages (VFTO), 209
Voltage escalation, 157, 160

Zero sequence components, 33